The Pyrocene

The publisher and the University of California Press Foundation gratefully acknowledge the generous support of the Ralph and Shirley Shapiro Endowment Fund in Environmental Studies.

The Pyrocene

HOW WE CREATED AN AGE OF FIRE,
AND WHAT HAPPENS NEXT

Stephen J. Pyne

UNIVERSITY OF CALIFORNIA PRESS

University of California Press
Oakland, California

© 2021 by Stephen J. Pyne
First paperback printing 2022
Library of Congress Cataloging-in-Publication Data
Names: Pyne, Stephen J., 1949– author.
Title: The pyrocene : how we created an age of fire, and what
   happens next / Stephen J. Pyne.
Description: Oakland, California : University of California Press, [2021] |
   Includes bibliographical references and index.
Identifiers: LCCN 2021001149 (print) | LCCN 2021001150 (ebook) |
   ISBN 9780520391635 (paperback) | ISBN 9780520383593 (ebook)
Subjects: LCSH: Fire—History—Social aspects. | Fire ecology. |
   Climatic changes—Effect of human beings on.
Classification: LCC GN416 .P86 2021 (print) | LCC GN416 (ebook) |
   DDC 577.2/4—dc23
LC record available at https://lccn.loc.gov/2021001149
LC ebook record available at https://lccn.loc.gov/2021001150

30   29   28   27   26   25   24   23   22
10   9   8   7   6   5   4   3   2   1

To SONJA
still burning bright

and

to LYDIA, MOLLY, KARLIE, ASHLEY, LINDSEY,
COLTEN, JULIE, IVY, and ESTHER
with the hope that they can find in the ashes
the sparks of a better world than was left to them

And upon earth he shewed thee his great fire;
and thou heardest his words out of the midst of the fire.

DEUTERONOMY 4:36

And they shall go out from one fire,
and another fire shall devour them.

EZEKIEL 15:7

Contents

Prologue

Between Three Fires

The fires seemed to be everywhere.

Places that commonly burn, such as Australia, California, and Siberia, burned with epic breadth and intensity. Australia had established a historic standard for a single outbreak with the 2009 Black Saturday fires; the 2019–20 Black Summer burns broke historic standards for a season. California endured its fourth year of serial conflagrations, each surpassing the record set the season before. Like a plague, the fires spread into Oregon and Washington, then leaped over the Continental Divide to scour the Colorado Rockies. The Siberian burns moved north of their home territory and flared beyond the Arctic Circle. Places that naturally wouldn't burn or would burn only in patches were burning widely. The Pantanal wetlands in central South America burned. Amazonia had its worst fire season in 20 years. And what the fires' flames didn't touch, their smoke plumes did. Australia's smoke circled the globe. The palls from the West Coast fires spread haze through half the country; they struck with the symbolic impact and visual intensity that dust storms evoked during the 1930s. The fires' smoke obscured subcontinents by day; their lights dappled continents at night, like a Milky Way of flame-stars. Where fires were not visible, the lights of cities and of gas flares were: combustion via the transubstantiation of coal and gas into electricity. To many observers, they appeared as the pilot flames of an advancing apocalypse. Even Greenland burned.

The smoke and flames were a symptom, not a syndrome. The planet's unhinged pyrogeography was also shaped by fires that should have been present and weren't. These were fires, historically set by nature or people, to which landscapes had adapted. Now those fires were gone, and the land responded by degrading ecologically while building up combustibles to stoke more savage wildfires. The Earth's fire crisis, that is, was not just about the bad burns that trash countryside and crash into towns. It was equally about the good fires that had vanished because they were actively extinguished or no longer lit. The Earth's biota is disintegrating as much by tame fire's absence as by feral fire's outbreaks. In 2013 the Pinchot Institute for Conservation surveyed the state and likely future of American forests. *Forest Conservation in the Anthropocene*—the outcome of its gathering of experts—included a full-body ecological CAT scan that looked at flora, water, air, soils, and wildlife. The one element every discipline included, the point of intersection among them all, was fire. Every aspect of the fast-morphing scene was touched by fire: it integrated everything else. If you didn't get fire right, you wouldn't get the rest right.[1]

There was a third facet to the planetary fire triangle, one that looked beyond present and absent fires to deep time. Its combustibles came not from living biomass, but from lithic ones. With increasing frenzy, humans were binge-burning fossil fuels. They were taking fuel out of the geologic past, burning it in the present with complex (and little understood) interactions, and then releasing the effluent into the geologic future. Industrial combustion restructured the dynamics of fire on Earth. Fossil fuel combustion acted, in brief, as an enabler, a performance enhancer, and a globalizer. It ensured that little of the Earth would be untouched by fire's reach if not its grasp.

The dialectic between burning living and lithic landscapes explains most of the paradoxes of Earth's fire scene. Paradox one: the more we try to remove fire from places that have coevolved with it, the more violently fire will return. Without the counterforce provided by petrol-powered machines, from helicopters to portable pumps, there could have been no serious effort to exclude fire in the first place. Paradox two: while wildfires gather more and more media attention, the amount of land actually burned overall is shrinking. Fossil fuel societies find surrogates for fire and remove it (or suppress it) from landscapes. California experienced 4.2 million burned acres in 2020; in preindustrial times, it would have known probably over 10 million acres burned, though not burned in wild surges. Paradox three: as we ratchet down fossil fuel burning, we'll have to ratchet up our burning of living landscapes. We have a fire deficit. We need to make firescapes more robust against what is coming, and fire may be the surest way to do it.[2]

Add up all these fire influences—those directly through flame and those indirectly through smoke, removed fire, fire-enabled land use, and a warming climate—and you have the contours of a planetary fire age, the fire-informed equivalent of an ice age. You have a Pyrocene.

What Is the Pyrocene?

Pyrocene proposes a fire-centric perspective on how humans continue to shape the Earth. It renames and redefines the

Anthropocene according to humanity's primary ecological signature, which is our ability to manipulate fire. It comes with a narrative—the long alliance between fire and humans. It proposes an analogy for the future—the sum of our fire practices is creating a fire age that is equivalent in stature to the ice ages of the Pleistocene. With fire as a theme, it offers a sideways view on climate change, the sixth extinction, changes in ocean chemistry and sea level, and the character of the human presence on Earth. It retells familiar stories from a different vantage point and introduces topics not previously considered fundamental. Like fire, the Pyrocene integrates its surroundings—geographic, historical, institutional, intellectual. It addresses the search for a usable future.

The history it tells chronicles three fires. First-fire is the fire of nature—fires that appeared as soon as plants colonized continents. Fossil charcoal traces their presence back 420 million years. Second-fire is fire set and abetted by humans. Thanks to cooking, a dependence on fire became coded into hominin DNA; thanks to favorable conditions at the end of the last ice age, second-fire steadily spread everywhere humans did. Together, they competed with first-fire and expanded the overall domain of burning such that very little of terrestrial Earth—places blanketed by ice, implacable deserts, sodden rainforests—lacked fire. Human-kindled fires burned as first-fires did, in living landscapes, subject to shared conditions and constraints. Third-fire is qualitatively different.

Third-fire burns lithic landscapes no longer bounded by such ecological limits as fuel, season, sun, or the rhythms of wetting and drying. The source of combustibles is essentially unbounded; the problem is the sinks, where to put all the effluent. Third-fire unhinged not only climate and biotas, but the affinity between people and fire. Second-fire was an act of domestication, perhaps the model for domesticating, in which people had transformed wild fire into hearth and torch just as they had cultivated teosinte into maize, and aurochs into dairy cows. Both fire and people spread in a kind of mutual assistance pact. There was a fundamental inequality in their relationship because fire could exist without humans while humans could not exist without fire. But each operated within common conditions.

Third-fire decoupled that relationship. People could exist without it, but it could not flourish without people. It was about power; not the power of fire to nudge, leverage, integrate, and quicken, but the brute force of fire distilled and mechanized. Second-fire was a kind of mutual taming, a partnership. Third-fire was just a tool. It generated raw power.

The three fires have competed, complemented, and colluded—an ecological three-body problem. But over the last century the terms of their interactions have changed. Something flipped. What had been a rheostat became a toggle switch. Earth's fire system crossed a summit into a new state, not easily reversed, as once-friendly fire morphed into feral flame. In unprecedented ways the Earth had too much bad fire, too little good fire, and too much combustion overall. It was not simply fire's indirect relationship to climate that was upset: the whole of fire's presence on Earth was unhinged. The sum of humanity's fire practices overwhelmed the existing arrangement of ecological baffles and barriers. Fire created the conditions for more fire. Unwittingly, humans had created a fire age, but whether they can inhabit that world is unclear.

It is a future that seems so dire, and its likely trajectory so strange, that some observers argue that the past has become irrelevant. We are, they fear, headed into a no-narrative, no-analogue tomorrow. So immense and unimaginable are the coming upheavals that the arc of inherited knowledge that joins past to future has broken. There is no precedent for what we are about to experience, no means by which to triangulate from accumulated human wisdom into a future unlike anything we have known before.

Yet the argument is misguided. Fire's past remains its prologue, offering both narrative and analogue. Where once there was one kind of fire on Earth, then two, there are now three. That's the narrative. Between them the three varieties of burning are sculpting a fire age equivalent in stature to the ice age of the Pleistocene. That's the analogue. Since the onset of the last interglacial, we have been fashioning piece by piece a more fire-friendly world that has eventually yielded a fire-informed one. Like fire itself, that world is assuming an autocatalytic character that makes more fire possible. Propagating ice previously helped push the planet into an ice age; likewise, our binge burning is now propelling the Earth into a fire age.

We have created a Pyrocene. Now we have to live in it.

1   *Fire Planet*

Fire Slow, Fire Fast, Fire Deep

Earth alone holds fire. It's worth pausing over this remarkable circumstance. Among planets fire is as rare as life, and for the same reason: fire on Earth is a creation of the living world. Life in the oceans gave Earth an oxygen atmosphere. Life on land gave Earth combustible hydrocarbons. As soon as plants rooted on land, lightning set them ablaze. They've been burning ever since.

Other planets have some oxygen—Mars, most notably. Others have combustibles; Saturn's moon Titan has a methane atmosphere. The gaseous planets have lightning. But none have all the necessary elements, or not in ways that allow them to combine. We may find exoplanets that have life, that perhaps will have fire, that may even have intelligent species who can manipulate fire. But we know of none now, and if we find some, they will be so remote as to offer no comparative value. We live on the only viable fire planet. If we were to visit another at some distant future, it is most likely we would do so on plumes of flame.

How the Earth Got Its Fire

Fire on Earth has a history. It has its narrative. There was a time when fire didn't exist, though it is difficult to imagine a time, short of the planet's immolation by an erupting sun, when it will no longer flourish. The Earth would have to lose its lands, shed the oxygen in its atmosphere, end the electrical imbalances between land and air that kindle lightning, and find another way to convert energy into terrestrial matter, and hydrocarbon molecules back into energy. All that is possible in principle. It just isn't plausible on this planet.

The history of fire, in brief, is the history of terrestrial life. Fire's evolution, or elaboration, into new varieties and expressions, its arrangement into ever-changing biomes, its reorganization into new pyrogeographies, its inextricable entwining with the other elements—all this is not a process parallel to the evolution of life, but a coexistence and even coevolution with it so shared that it comes close to symbiosis. Fire is not alive, but because life called it into being and sustains it, it shares, much as a virus does, many of life's properties. It feeds on living biomass, it spreads by a contagion of combustion. *Fire of life* is not a random metaphor.

Because it is a reaction, not a substance, fire is what its setting makes it. Fire's history is the history of its parts and how they come together. Like a driverless car, it has no single pair of hands on the wheel: it synthesizes its surroundings. It takes its character from its context. It barrels down the road integrating everything around it. As air, water, earth, ignition, and life change over time, so does fire.

Begin with the oldest component, lightning. In truth, the Earth has plenty of sparks, and fires have started from rock falls, earth slides, volcanoes, spontaneous combustion, and the occasional meteor, but only lightning can account for the prevalence of combustion on a planetary scale. It has been continuous since the early Earth. Its relentless pervasiveness is a major reason why fire is elemental, ancient, and inescapable.

Lightning can seem capricious. It is not evenly spread over the Earth or across eons: it appears in clusters, crowding in time, bunching in space, tethered to places amenable to thunderstorms. Only a fraction of lightning is capable of kindling fires—only bolts that connect land to cloud, not cloud to cloud; bolts that strike something combustible, not rocky peaks or lakes; bolts with the proper electrical properties, throbbing and full of heat.

Both fire and lightning highlight a complex choreography between wet and dry. Some moisture is needed to make storms, and so lightning; too much moisture prevents ignition. Some moisture is needed to grow fuels; too much renders fuel unburnable, and too little means fire can't spread. Areas dense with thunderstorms rarely equate with areas rife with lightning-kindled fires (central Florida is a notable exception). Dry lightning, in which rain evaporates or is separated from the bolts, starts more fires than wet lightning, whose spark must survive a downpour.

Compared with the number of flashes, fires are relatively rare, but when conditions are right, they can kindle fires in swarms. The epicenter of lightning fires in the United States is the Southwest, where drought and monsoon, mountain and desert, make ideal circumstances for dry lightning. Yet outbreaks can come even to places that experience them rarely. In Northern California, the 1987 fire bust ignited 4,161 fires, of which 92 grew larger than 300 acres. In the siege of 2008, lightning started almost 3,600 fires, of which 88 swelled over 1,000 acres. In 2020, in excess of 10,000 strikes were documented amidst a record heat wave, and they kindled between 400 and 500 fires, most around the Coast Range.[1]

The interaction between life and lightning is unequal. Lightning is a phenomenon of geophysics, not of life. Plants adapt to lightning, lightning does not adapt to plants. Lightning can occur on Jupiter or Uranus as readily as Earth; a bolt can select a limestone cliff as handily as a black spruce. There seem few ways for the living world to influence the electrical discharge except very indirectly, as seen when taller trees are struck more than short ones. Lightning can exist without life; fire cannot, and so it shares life's evolution. If life were removed altogether, lightning would continue. Fire would expire.

Fire needs more than ignition, and life contributed the two other ingredients—oxygen and fuel—that stir together to make combustion and, in the right context, fire. Life in the oceans first stuffed the atmosphere with oxygen, and life on land lathered the surface with fuels. When those products of the living world met lightning's spark, fire resulted. It is not just that planets like Mars lack life because they lack the proper conditions, but that they lack those conditions because they lack life.

The varieties of organisms that captured and tamed oxygen came late in the process. The earliest forms of life (in oceans) appeared in the absence of oxygen; the first photosynthesizers were anaerobic; many genera today still thrive in anaerobic settings in swamps, lakes, and seas below an oxygenated zone, and there are whole ecosystems in the deep oceans around thermal vents that operate on a fundamentally different chemistry. For all of these biomes, free oxygen could prove toxic.

The dominant life-forms of the time, however, managed to turn oxygen from a threat into an opportunity. They captured it, contained it, trained it to their purposes—a template for what hundreds of millions of years later hominins would do with fire. They didn't merely accept oxygen, they produced it. A competition commenced between organic sources and geologic sinks. While life evolved new ways to release oxygen, rocks found new ways to absorb it. Then between 2.35 billion years ago and 700 million years ago, during what has become known as the Great Oxidation Event, sources overwhelmed sinks, as atmospheric oxygen increased and stabilized, and aerobic photosynthesis and respiration became the norm for terrestrial life. What had been a chemical poison evolved into a biochemical necessity. Ever after, across the swells and troughs of deep time, the Earth's atmosphere has contained oxygen in quantities that range between 14 and 16 percent, below which it is difficult to get biomass to burn, and 30 and 35 percent, at which it is hard to halt burning.[2]

A new process, combustion, became possible, then pervasive. Note that its chemistry is a *bio*chemistry: it occurs within living organisms and by means created by life. Within cells everything about the reaction is tightly constrained to prevent free-roaming oxygen from wreaking havoc. Across landscapes, however, little is confined, and fire responds to winds and humidity, to seasons and rains and droughts, to terrain that ranges from gorges to mountains, and to infinitely complex arrangements of biomass over which humans have little power.

Life on land had to find ways to control fire, or at least influence its character, or everything that grew might be scavenged by burning. What happened with oxygen thus happened with fire: what began as a potential toxin became a norm, and then a necessity, and life even moved to enhance its presence. Terrestrial life and fire coevolved within a shared biotic matrix; they became interdependent and, in curious ways, codependent. In brief, fire was not something imposed on life, like wind or flood. It emerged from life's very character.

Earth's is an atmosphere of combustion, primed to burn, but also an outcome of combustion. As James Lovelock memorably put it, "It is not too far-fetched to look on the air as like the gas mixture that enters the intake of an internal combustion engine: combustible gases, hydrocarbons, and oxygen mixed. The atmospheres of Mars and Venus are like the exhaust gas, all energy spent." The nature of this interdependence is tricky, and an inquiry might usefully begin with the question—a template for all the others that fire seems to raise—of whether fire has influenced the atmosphere within which it burns. How does the fast combustion of fire compare to slow combustion of respiration in shaping the planet? Is free-burning fire a vital process in the global oxygen cycle, or simply a geochemical afterthought? By regulating carbon, has fire also regulated oxygen? Geologic eras with more fossil charcoal suggest higher oxygen; those with scanty charcoal hint at lower. As oxygen rises and falls, so does the planetary condition for fast combustion.[3]

What was the character of burning across deep time? Surely, the Earth burned in the past, as it does today, with a medley of fires, some of which flare at certain times and dim at others, but all of which depend on the character of the biomass available to burn. Oxygen content would have to rise significantly for wet logs to burn, and it would have to drop hugely to prevent ignition in small, dry grasses. By whatever feedback fire shapes the atmosphere, it seems to do so through fuels. After all, the photosynthesizing plants that pump oxygen into the air are the same ones that stoke free-burning fire.[4]

It doesn't take long before the question begins to turn on itself like a Möbius strip. And it's not just oxygen. There is a perhaps unavoidable circularity to fire that may not be inappropriate for a phenomenon whose very character is to interact and integrate.

The Fire of Life

Fire's biological makeup is fundamental. It takes apart what photosynthesis puts together. When that occurs in cells, we term it respiration; oxygenation occurs among tightly scripted and bounded molecules. Call the process slow combustion. When it occurs in the wide world, we label it fire; oxygenation takes place amid an essentially unbounded milieu of roughened terrain, turbulent air masses, and an endlessly evolving biota. Call it fast combustion. These processes have gone on since the Devonian, over 420 million years. Call that chronicle deep combustion. Fire slow, fire fast, fire deep—they are as basic to Earth as the waters that flow through its valleys and the plants that clothe its slopes.

This vision can seem bizarre to people who live in cities in the industrial world and whose fire power resides in machines. They experience fire mostly virtually, through monitors. They no longer use flame to do routine work—to light scenes, to cook food, to heat rooms, to jolt fields and pastures to life, to protect themselves against wildfire. There are practical reasons, of course, to restrict flame in built landscapes, but there are cultural biases as well. European thinkers have long regarded the use of fire as a stigma of primitivism, while European agronomists identified agricultural burning with prerational superstition. Even the term *developed country* refers, in shorthand, to a nation that has replaced open flame with the closed combustion encased in fire machinery. Out of sight, out of mind.

If such residents can imagine fire, they see it as a solely physical process. They define fire as the oxygenation of hydrocarbons, a chemical event shaped by its physical surroundings. They imagine it as a phenomenon that physics and chemistry can deconstruct and then implant into devices like stoves, candles, and furnaces. They can have light without heat, heat without smoke, smoke without fire, fire without flame. They can live in a setting that may burn but only as accident or arson and almost always as disaster. They have broken down fire into its components and refashioned them to support a world of combustion without fire. If they think about fire on the land, they likely imagine it as another spasm of physical force like a tornado, tidal wave, or flood. To them it is something imposed on landscapes from outside; an ecosystem may adapt to it, as a river's channel to flooding, but life has no say in that jolt of energy any more than it does with an erupting volcano or an earthquake.

Yet fire is intrinsically different from these other disturbances. It emerges out of the character of living landscapes. In the living landscape all the factors that literally feed fire converge. While ecological science formally regards fire as a disturbance, this is a fiction of modeling, for fire is a "disturbance" in the same way rain is. Even fire's chemistry is a *bio*chemistry. Hurricanes and floods can occur apart from anything living. Fire cannot. It derives its power by feeding on its biotic matrix. It more resembles a fast herbivory than a wind or ice storm. Like a virus, fire is not itself alive, but it relies on the living world to propagate. We often speak of an epidemic spreading like wildfire, but it makes equal sense to speak of a fire spreading like a plague, a contagion of combustion.

. . .

Ignition in the form of lightning is the oldest contributor to the reaction we call fire, but it is the patchiest in any particular place. Oxygen arrived next, but it is universal—the same around the globe for any particular era. The final ingredient is fuel. Life had to leave the oceans in order to burn, and when it did, combustion found a new abode in which to happen. Combustion could spread between organisms; fast combustion, fire, could complement and compete with the slow combustion of respiration. The terrestrial biosphere has three ways to decompose biomass: microorganisms, herbivores, and fire. All three rely on forms of combustion. In the living landscape all the factors that literally feed fire converge.[5]

The complexity of combustion increased exponentially. Oxygenation took place not only as slow combustion among tightly scripted and bounded molecules in cells, but as fast combustion amid an essentially unbounded milieu of roughened terrain, turbulent air masses, and an endlessly evolving biota. Fire synthesized that setting: it took on the character of everything around it. Those interacting parts, not any single factor, made fire.

Fire changed as life changed. It evolved as producers, consumers, decomposers, predators and preys, grazers and grasses appeared and vanished, as life brushed against extinction and exploded into novel forms. Fire burned one way amid organic soils, another amid woodland savannas, still another through dense shrublands and gregarious conifer canopies. It furnished a fiery index for ecological conditions and evolutionary history. Those species that could not accept this fact, like those anaerobes that could not accommodate oxygen and retired to anaerobic niches, were doomed to occupy the incombustible environments of the Earth.

Yet many places and periods probably did escape fire. Fire is a particularized event: it appears patchily in space and time. Not all biomass is available as fuel; not all available fuels exist at the right time and place for lightning to kindle. It would be easy for a particular place to miss fire for a while. It would be unimaginable, however, for it to vanish from the Earth overall.[6]

What makes the biotic character of fire particularly fascinating is that fire became itself a process for evolutionary selection. Plants could accommodate its presence, and their evolutionary adaptations could then affect the character of fire. Stated so baldly, the argument is too simple. Organisms don't "adapt to fire" any more than they "adapt to rain." In each case they adapt to patterns of fire and water. A tree that can flourish in a regimen of precipitation that falls evenly every month is not adapted to a regimen that crowds an equal amount into 3 months. So, too, a tree adapted to a fire regime of near-annual surface burns cannot tolerate fires blasting through its canopy. The fire regime is a statistical composite, like climate; what storms are to climates, fires are to fire regimes.[7]

The spectrum of fire adaptations mirrors the complexity of living landscapes. Few adaptations are unique to fire; organisms display a suite of traits that adapt against a suite of stresses. Fire has been around so long—all the evolutionary history of terrestrial life—that its accommodation has merged with many others. What is doubly intriguing is that those adaptations can trend two ways. In the first, some species adapt by protecting themselves from flames, for example by thick bark, by flowers and seeds shielded by dense leaves, or by vital reserves stored underground.

A classic example is found in prairie. Putting most of its biomass belowground can help a big bluestem, for example, withstand drought and grazing as well as fire, and since fire occurs during dry spells, and grazers are attracted to the fresh growth that follows after a burn, a shared adaptation accommodates all three stresses. (The preference for new growth is as ancient as it is common. The fossilized stomach contents of a nodosaur, an herbivorous variety of ankylosaur, include charcoal, suggesting that 110 million years ago the ecology of fire and browsing was well established.)[8]

Yet in the opposite trend, some features do seem specific to fire, and some species appear to rely on burning for competitive advantage. Like chamise, a shrub in California chaparral, they enhance their flammability by promoting small leaves and twigs, boosting the proportion of dead to live particles, adjusting seasonal phenology to times of dry lightning or high winds, and amplifying chemical composition to favor combustion. So, too, serotiny—closed cones that release seeds only when heated, most often by flames, such that seeds drop down into an ash bed free of competitors—is hard to reconcile with other challenges a plant confronts. Yet it is common among boreal conifers like black spruce and jack pine and is found among proteas in South African fynbos. Promoting chemical flammability or depending on crown fires to propagate is hard to understand other than as a paradoxical accommodation to fire's inevitable presence.

Such traits can seem counterintuitive, especially to people who are removed from the routine use of fire and whose habitations are designed to abolish fire. Modern cities design structures to evict fire, not encourage it. The living landscape is different. To burn more briskly than other plants, to burn earlier than others around it, to grow gregariously in ways that allow fire to burn continuously, to help incinerate a biome—all may enhance a plant's own prospects if it can use heat and smoke to stimulate flowering, to regenerate more quickly from the ash, or to flourish more robustly in the postburn environment than its competitors. Such species need fire within an expected span of time. They—and the entire ecosystem that they shape—may actually suffer from a lack of fire.

Fire is as complicated as its setting: similar fires can have different origins. In nineteenth-century America, megafires broke out because logging slash lathered landscapes over large areas and human ignitions littered the countryside; in twenty-first-century

America, megafires respond to global warming acting on landscapes typically stuffed with combustibles from decades of fire's exclusion. Similarly, there are multiple paths to a given regimen of burning, and a common pattern of ignition (say, people burning along routes of travel) can express itself differently in boreal forest, prairie, tropical savanna, or shrubland.

What, exactly, does fire do? Its actions are both precise and generic. It shakes and bakes. It deconstructs biomass and readies the site for a new reassembly of materials liberated by burning. Around its flames revolves an ecological triangle, a circulation of biochemicals, species, and communities. It stirs molecules, organisms, landscapes. It kills plants, breaks down ecological structures, sets molecules adrift, shuffles species, opens up niches, and for a time rewires the flow of energy and nutrients. Fire upsets, quickens, shreds, reorganizes, and rejuvenates. It is at once both radical and conservative—radical in smashing an existing order, conservative in stimulating the conditions for the return of that order. It is, for nature's economy, the superlative example of creative destruction.[9]

Fire occurs not only within a physical context of rock and wind and water, but within a biotic matrix. The point of engagement where fuel, oxygen, and spark converge is the living landscape. Biomass integrates much of the outside world; fire integrates combusting biomass with everything else. Fire feeds on it, sculpts it, consumes it, catalyzes it, and depends utterly upon it. By means both coarse and delicate, fire can shape the world that shapes it.

.   .   .

Across the centuries fire ceremonies have tended to celebrate the ability of fire, especially fire in the hands of humans, to purge the bad and promote the good. Witches have been burned, cattle have passed through smoke to be cleansed of parasites, young couples have leapt over flames to encourage fertility. This same perception underwrites much of agriculture outside floodplains, which is an exercise in applied fire ecology, relying on fire's capacity to fumigate and fertilize, to drive off microbes and competitor plants while transmuting otherwise inaccessible lumps of hydrocarbons into available nutrients. Yet an appreciation for fire's ecological presence has come haltingly. Not until the late 1950s did fire ecology have a name, and not for another decade, the sketchy contours of a discipline.

Because the prior science condemned fire as damaging and primitive, few researchers had seriously examined fire's complex character. Rather, they documented the ways it harmed what they valued, like mature timber and organic soil, and they sought means to understand how it behaved in order to better control it. Their critics, grounded in fire's use rather than its suppression, began by recording fire in traditional, cultural landscapes, where burning was a way of life, not simply an ecological process. They examined the ways fire interacted with fire-tolerant and fire-promoting pines, hickory and oak savannas, tallgrass prairies, talismanic sequoias. They observed the malign consequences of removing fire from biomes that had adapted to it. They amassed enough data to challenge prevailing management paradigms that had sought to exclude fire. All this occurred in familiar landscapes and along familiar themes. In the end much of what resulted flipped prior understanding.

Slowly, researchers began probing the domains of fire more widely. They found that smoke was not simply a nuisance, but an active ingredient in the atmosphere and a stimulant for many plants. They discovered that plumes carried the agents of microbial dispersion, not unlike ocean currents transporting species across the seas; they realized that smoke could enhance photosynthesis by dispersing sunlight to leaves other than those directly exposed; they saw how late-fall smoke, by cooling streams, could assist salmon migration; they saw how smoke prompted pineapples and grass trees to flower, and South African veld grasses to thrive more vigorously than those not smoked. They learned that charcoal was not simply a by-product to be blown by the winds and swept by rains, but a component, often critical, of soils. Rather than simply volatilizing nutrients and stripping soils, fire could, through biochar, enrich a site's suitability for plants—as it did in Amazonian lands laden with biochar, which flourished with higher productivity than char-barren patches. They saw how the mixed severity of fires and the dappled geography of large-area burns promoted biodiversity. They saw geotropic orchids that flourished around Cape Town within 48 hours of a burn, and how lodgepole and jack pine in the northern Rockies reseeded massively after searing crown fires. They found that postburn erosion into rivers was a major source of carbon deposition in oceans. They understood how fire contributed to the planetary cycling of carbon. More and more, fire research confirms what those ancient ceremonies celebrated. Fire tears down and builds up. It recycles and rejuvenates. It jolts out of biotic lethargy—a broad-spectrum ecological catalyst. Our appreciation for how widely fire interacts with Earth's biota seems limited only by our willingness to look.[10]

Definitions depend on purpose. For agencies that must manage fire, a useful categorization is to consider biomes as fire dependent, fire sensitive, or fire independent. Fire-dependent biomes are well adapted to fire, are maintained by fire, and suffer if fire is lost. Fire-sensitive biomes have not evolved with fire, show no distinctive adaptations to it, and can be harmed by it (which is typically a product of human finagling). Fire-independent biomes accept fire—are fire tolerant—but their persistence does not require it. Sometimes a fourth category, fire influenced, is introduced to characterize ecosystems that, mostly due to human activities, seem to be shifting between fire-sensitive and fire-dependent status.[11]

In 2002 a Global Fire Partnership between The Nature Conservancy, the World Wide Fund for Nature, and the IUCN-World Conservation Union used these categories and assessed 46 percent of global ecoregions as fire dependent, 36 percent as fire sensitive, and 18 percent as fire independent. Overall it reckoned that 84 percent of major habitat types had degraded fire regimes—too much fire, too little, or the wrong kind. Additionally, an uptick in fire research in recent decades has tended to further shift what had been regarded as fire sensitive (for example, oak-hickory forests) to fire influenced, and it may be shifting it to fire dependent. With humanity's increasing meddlings, we can expect more reclassifications.[12]

Fire is nature as postmodernist: it is all about context. Meanwhile, the more we search for fire's presence, the more we find. The old order of understanding demanded explanations for fire's presence; the new order, for its absence.

The Paleohistory of Fire

Earth had a long primordial period without fire; then, as life emerged in the oceans, a period with combustion but not fire. Sometime between 450 and 420 million years ago (MYA), as life clambered onto land, the pieces snapped together with enough force to burn and keep burning. Combustion evolved into fire. Before that moment, fire did not exist; afterwards, it became almost impossible for it not to.[13]

The first fossil evidence for fire appeared in the Silurian, around 420 MYA, about the time of the earliest vascular plants. Through the Devonian (419–359 MYA), plants diversified, and the first forests appeared. For a while charcoal disappears from the sedimentary chronicle, perhaps from the relative thinness of fuel, more plausibly from the low level of oxygen (15–17 percent). The geologic record, however, is full of gaps called unconformities where erosion has removed the record of the rocks, and the sparsity of fossil charcoal (fusain) may simply be the unconformities of fire history. Still, Earth remained a world of increasing vegetation. Around 375 MYA the Earth slid into an era of global cooling, slowly flipping from a greenhouse to an icehouse climate. Biomes moved; fire followed. Overall, however, this was a low-fire world, limited by the availability of stuff to burn.

By 350 MYA, despite a continued icehouse climate, increasing oxygen and denser vegetation, where it was possible, announced the onset of a high-fire world fed by enriched oxygen and thickening fuels. Fires spread; fire's realm widened and deepened. What were those ancient fires like? Simply put, they were like the fuels on which they fed. They burned through conifers and angiosperms, they burned into tropical mires, they burned across upland shrubs and woods. Especially they swelled and sagged with the disposition of the available fine fuels. But what were the fine fuels of the Pennsylvanian, or the Jurassic? Long-needled gymnosperms did not evolve until the late Paleozoic, deciduous angiosperms until the Late Cretaceous, and grasses until the Miocene. The atmosphere's high oxygen content allowed insects to swell to gargantuan sizes—dragonflies the size of possums. The same likely holds for its fires.

The geologic record favors the big and the tough over the tiny and the easily volatilized; the best preserved fine fuels are those in fossil charcoal, charred rather than burned. As with carbonized scrolls from ancient Pompeii, new techniques allow more and more of that record to be read. Early fires burned amid reedlike psilophytes and pteridophytes, within once-sodden swamps of rotting debris, biomes later enriched with horsetails, wood and soft-leaved ferns, towering lycopods, and *Calamites* trees with branches that whorled like maypoles. All could combust under the proper conditions, and some perhaps could sustain spreading fire fronts. Analogues that exist today burn nicely: while palms shrug off fires like raindrops, bracken fern carries flame with the wind, and swamps, drained by drought, readily refill with smoldering ash. Yet, although they undoubtedly combusted, such ancient biotas probably bear no more relation to recognizable wildfires of today than do lepidodendroids to lodgepole pine, or psilophytes to tallgrass prairie.[14]

Though much was burned, much was also buried. Fossil biomass fills vast strata of the sedimentary record, stuffing, for example, the coal beds of the Mississippian and Pennsylvanian periods, among others. Yet here, too, the chronicle testifies to fire. Fusain fills 2 to 13 percent of the coal beds of the Pennsylvanian. (In marine deposits of the North Sea, charcoal is regarded as "the most common form of fossil plant preservation.") The burning must have shaped those biomes as much as those biomes shaped fire. Still, by today's standards, the fire cycle was out of balance. Fuel sources far exceeded fire sinks, producers raced ahead of decomposers, and fuels piled up faster than fires could remove them. It is not impossible that a creature evolved from among evolution's menagerie that could kindle fire and act as a broker to balance fuel and flame, or at least distribute it as some raptors do today in northern Australia savannas, though a torch-wielding velociraptor would make for a truly terrible lizard. But there is no evidence as yet to suggest it happened.

The serious interruptions in fire's chronicle map onto mass extinctions, which affected both fuel and oxygen. At 250 MYA, the Permian-Triassic boundary, perhaps 90 percent of life ended, marine as well as terrestrial. Life nearly died, and fire along with it. Fire does not reappear at strength until the Late Triassic. It all crashed again at the Triassic-Jurassic boundary (200 MYA), then recovered with abundant fires burning through a flush suite of biomes. The Cretaceous was a high-fire world with a greenhouse climate. Oxygen was probably high, flora witnessed the spread of angiosperms (flowering plants), polar ice was gone. There were fern prairies, coniferous forests, deciduous woods, dinosaurs, plenty of fossil charcoal. The era ended with another extinction event that marks the transition to the Tertiary (65 MYA), an event best known for the meteoric impact at Chicxulub. How much of a burn the bang kindled is unclear. There were flash fires as particles of the disintegrating meteor fell like hail, but the planetary trauma left lots of future fuel in its wake—a cosmic example of slash and burn. So while fires continued across the Cretaceous–Tertiary (K-T) extinction event—fusain says as much—they must have changed their character, just as flora and fauna did. Dinosaur fires were replaced with mammal fires, as it were.[15]

By the Tertiary period, the fusain fraction in sediment has fallen to less than 1 percent. Oxygen has stabilized, approaching modern values—no longer a primary consideration in the variables that organize fire. (Today's fires burn in an atmosphere less chemically active for combustion than at any time since the Carboniferous.)

Climate and biotic evolution assume prominence. Yet the Paleocene-Eocene Thermal Maximum, an intense spasm of global warming that occurred around 55 MYA, did not leave a rich record of charcoal; whether something dampened burning, or the evidence of burning has not yet been discovered, is unclear. But the apparent absence of fire is noteworthy.[16]

For Earth as a fire planet, two events from the era stand out, representing the planet's pyric polarities. One is the emergence of tropical rainforests, not a realm for routine fire. The other is the emergence of grasses, most of which readily support fire. Of particular note is the appearance of savannas composed of C_4 grasses, capable of using carbon dioxide more efficiently (and at lower atmospheric levels) in photosynthesis and also exceptionally prone to burn. Where they establish themselves, a "coupled fire-grass system" thrives and often dominates. Grass replaces the classic sources of charcoal. The appearance of new biomes is a powerful reminder that fire is not solely the result of physical factors, like climate, oxygen, carbon dioxide, and inert hydrocarbons, but of how life responds and how its inventions can feed back into the overall system. The arrival of a new organism (like cheatgrass or *Imperator*) can restructure a fire regime independently of oxygen and climate.

In short, fire history parallels the general geologic history of Earth. The eccentric ecology of fire evolved along with the oft-lurching evolution of its parts. While fire's core chemistry persisted unchanged, its expression in landscapes shape-shifted with the idiographic histories of oxygen, climate, flora, and fauna. Its vigor rose and fell with the swells and troughs of oxygen. Its geography shuttled between icehouse and greenhouse climates. It collapsed with crashes—the great extinction events that wiped out or reorganized terrestrial life. Life's accommodations to fire nudged and morphed throughout. Charcoal, which is virtually immune to further decomposition, holds the archives. Fire was the means to preserve the record of its own history. It was both actor and archivist.

This brings the tale to the last 2.6 million years. It's a time of planetary cooling—a chill that began in the Pliocene. But combined with other factors, the global cooling reinstated an icehouse world and led to the whipsaw frost-thaw cycles that define the Pleistocene. The Pleistocene announced another—the fifth—global extinction and a time, being mostly wet and cold, that dampened fire. Yet again biological evolution proposed a counterforce. The hominins arrived, acquired the capacity to manipulate fire, and when conditions were right, notably as the last glacial maximum waned, they began redefining the role of fire on Earth, which meant remaking Earth itself.

. . .

It's an epic history, but one documented at scales both majestic and minuscule, and largely through proxies. The known plants are mostly those preserved in shales and coals. The carbon dioxide and oxygen levels are those extrapolated from carbonate rocks and charcoal. There is nothing peculiar to fire history about this: it is the nature of geologic history overall, which must deal with rocks from specific sites, amid unconformities that have erased vast amounts of time, and with extrapolations from relationships that allow something like oxygen levels to be guessed through their effects on combustion. That we know anything about the geologic history of fire is astonishing.

A few observations seem to stand out. One is the antiquity and essential naturalness of fire. As soon as the conditions existed that made it possible, fire appeared, and as those conditions have evolved, so has fire. Much as Earth has shuffled between icehouse and greenhouse climates, so it has moved into and out of high- and low-fire worlds. Some past fires have gone extinct, along with pterodactyls and aptosaurs, but fast combustion has been continuous.

It also seems apparent that a biological chasm could exist, as it did in the coal beds of the Paleozoic, between what might burn and what did burn, as manifest by the abundance of both burned and buried biomass. Perhaps the biomass was simply unavailable—matter, not fuel—because the climate lacked the proper wet-dry rhythms to crack open and dry out the vegetation, or because the right animals did not exist to munch and crunch biomes into burnable states, or because so much lay in swampy environs beyond oxidizing (and fire). A no less likely explanation is that ignition was too random and fire too geographically specialized. Fuels could hide in wet nooks and seasonal crannies from the predatory flames. The Earth lacked a fire broker, a creature capable of reconciling fuel's supply and flame's demand. The epoch's colossal stockpiling of carbon tracks a fire deficit so vast that it had implications for the global climate, as its fire-catalyzed release in recent times does today.

The simplest explanation may be that not all the elements of nature's fire had been brought under biological control. Slow combustion long ago impressed itself into the DNA of Earthly life. And then life began to command the conditions that made fast combustion possible. Until that happened, fire could synthesize fuel and oxygen only spottily, if brilliantly, veering between fires huge and tiny. Those historic gaps ended with the arrival of fire-wielding hominins, who first made ignition as steadfast as air and then, with the dominance of the sapiens, readied fuels for the flame and, in fact, did not limit their quest to fuel foraging but planted and slashed what they wanted to burn and even ripped additional combustibles out of ancient rocks.

The dominant story today is the reverse. Although carbon continues to recycle—charcoal from free-burning fire is one of the few mechanisms for shifting black carbon from the biosphere directly into the lithosphere—carbon is not being stored but released. Humans have exhumed fossil biomass and are burning it on such an immense scale that combustion and fire regimes now extend across geologic time. What failed to burn in the ancient Earth is burning in the modern. The limits of fire have increasingly become only those imposed by human will.[17]

The Enlightenment: Fire's Dark Age

For over 2 million years hominins have used fire—could not exist without it. They flourished where fire was possible and struggled in stony deserts, soaking rainforests, and shade forests that lacked wet-dry seasons where it was not. Their fire lore was embedded in social roles, codes, laws, rituals, ceremonies, and practices. They had fire in their shelters, their communities, their fields, pastures, hunting grounds, berry-harvesting patches, their paths and songlines. They used it to wage war and to celebrate peace. Throughout, they managed to coexist with fire—birthed it, tended it, trained it, related myths about it, and told almost all stories around it.

Then, in temperate Europe, with the advent of the Enlightenment's modern science and the technologies it made possible—if we want a date, pick Antoine Lavoisier's discovery of oxygen—hominins began to lose their fire connection. Fire ceased to be a phenomenon that saturated the world and human society—ceased to be a relationship—and became simply a tool, an implement to advance human power. Social elites decided that fire had to be deconstructed, put into appliances, or as coal became a dominant fuel, sublimated into steam. Open burning reeked of superstition and magic; it was deemed dangerous and unnecessary. Science would show how to create substitutes for it or, where that was difficult, to simply suppress it; and the same determination was applied to traditional knowledge. Fire went from friend to foe. Combustion became once more an informing principle, but this time as an unbounded source of anthropogenic power by which to disrupt the Earth. The blowback has been colossal.

Of all fire's stories the account of Europe's understanding is one of the most curious. Mediterranean Europe had appreciated fire— had made it one of four elements, had designated two of the twelve Olympian gods (Vesta and Vulcan) to honor it, had made it a central question for science and philosophy. Folk used it widely, though officials and agronomists proved wary. But with a Mediterranean climate that had an annual cycle of wetting and drying, and that subject to episodic droughts, fire would thrive in the living land-scape. It could be tamed, but not denied. In contrast, northern—temperate—Europe had no natural basis for fire, no regular wet-dry rhythms, no dry lightning, no dry mountain winds. The region was a great anomaly on a fire planet. There was fire aplenty, but only because people put it on the land. Folk used it—had to use it to make the land habitable—but elites concerned with modernizing econo-mies viewed it with suspicion, came to disdain it, and pined for an alternative. The best known of northern Europe's fire gods was Loki, an irredeemable trickster.

This matters because the nations of temperate Europe became the center for modern science, led the second wave of European imperialism that spilled out in the eighteenth and nineteenth centuries, and forged a new kind of combustion, the burning of fossil fuels. They wanted fire's utility without its froward, shape-shifting character. They wanted it harnessed in a machine—wanted to extract its specific powers without the burdens and messiness of foraging for fuel, breathing smoke, and endless tending. Elites regarded its comparative use as an index of social order and progress—in engines and furnaces, rational and good; in fields and pastures, superstitious and slovenly. With industrial combustion as an alternative, they considered fire a relic of barbarism and considered success at devising an alternative a mark of Reason. Like most peoples, those of temperate Europe took themselves and their landscapes as normative. Elites regarded Enlightened science as the supreme form of knowing and took their new industrial power as an ideal surrogate for fire and the triumph of physics (and reductionism) over biology (and synthesis). Unlike most peoples, temperate Europeans were in a position to influence fire's history across the globe.[18]

In 1848 Michael Faraday could use a candle to illustrate all the principles of modern science, confident that his audience could relate to his examples. But working fires were disappearing; coal, then oil, replaced wood and wax as a source of motive power and illumination. A century later the only university fire department was one that dispatched emergency vehicles when an alarm sounded. As fire ebbed from daily life, elites actively sought to remove it from living landscapes, and when fusain was discovered in coal beds, they first debated its character, then insisted (despite Charles Lyell's support) that it was not fossil charcoal, and so removed fire from lithic landscapes and geologic history. Fire found itself extinguished from both lived experience and historical record. Fire seemed the biological equivalent of a typo or an act of social vandalism. It survived as disaster and a register of social messiness.

At immense labor and expense, fire was removed not only from cities and fields but from nature preserves and wildlands.[19]

Not until the late 1950s did it return. In 1958 Tom Harris published a paper describing fossil charcoal from Mesozoic strata in England. That same year the privately endowed Tall Timbers Research Station received its charter to investigate "fire-type" landscapes. A year later Harold Lutz published an account of fire in Alaska that admitted the prevalence of lightning as an ignition source (how educated observers could have so misread the fire scene for so long staggers the imagination). Both Harris's paper and the Tall Timbers charter lay outside established science—Tall Timbers refused to submit the papers of its annual conference proceedings to peer review, on the grounds that letting any established fire scientist, most all of them foresters, review a paper was the same as censoring it. Not until the 1960s was fire ecology granted a name; not until the late 1970s was the antiquity and ubiquity of fire recognized and the effort to expunge it wholesale admitted as a blunder; not until the twenty-first century, with the advent of megafires and the blatancy of climate change, has fire stormed back to reclaim the inquiry its presence deserved, and still its many avatars lack a shared organizing principle. Part of the enterprise to restore good fire has been to recover its evolutionary and cultural history.

It makes for a weird saga. Perhaps the best explanation is one William James offered, that "articulate reasons are cogent for us only when our inarticulate feelings of reality have already been impressed in favor of the same conclusion. . . . The unreasoned and immediate assurance is the deep thing in us, the reasoned argument is but a surface exhibition. Instinct leads, intelligence does but follow." Fire in the felt life of hunters, foragers, farmers, and pastoralists ensured that they continued to imagine it in stories and explanations of the world. By the same token, it also meant that northern European intellectuals, even (maybe especially) those passing through the transmuting flames of the Enlightenment, believed fire was far from natural, because temperate Europe had few natural fires, and those freakish. Revealingly, Europe's classic epigram on fire was expressed in terms of utility and social relationships: fire made a good servant but a bad master. Fire was an emblem and instigator of social derangement that was best caged and ultimately replaced.[20]

All this matters because the legacy of these traditions endures. Unlike the other ancient elements, fire still has no academic department of its own, a homeless subject sleeping rough in the streets and wandering among various disciplinary shelters. Its fundamentally biological character remains elusive, subordinated to physical models, much as the controlled combustion of mechanical engineering has encased it in machines. Even those who use it consider it a tool, like a candle or a turbine, not appreciating that when a drip torch puts flame onto land, the fire ceases to be a physical device and metamorphoses into an ecological process. Today the urbanites of industrialized societies have little personal contact with fire. It remains remote from their lived experience. They know it only virtually through various media, where it appears as disaster and disorder, or through its smoke, which poses threats to public health.

Even the language we use betrays fire's awkward standing. Fire is a source of metaphors, rarely a recipient. We say that an epidemic spreads like wildfire, while it might be more appropriate to say of a fire that it propagates like a disease, that its embers resemble a swarm of locusts, that its flames feed and its convective columns breathe, that it prowls through the landscape of Earth history as a bear might search out berries, grubs, and fish, roaming or hibernating with the seasons, growing fat and thin with the yearly offerings. The selective, patchy character of burning, its uneven expressions and what often seem like random outcomes, might be better expressed by viewing fire through a biological prism rather than a physical one.

As astonishing as are contemporary researches into fire's ecology and revelations about its primordial past, more astonishing still is what was once known about fire and has been lost. When we shifted to burning fossil biomass, the shock caused cultural amnesia about our heritage of fire. We replaced or suppressed our traditional knowledge, and our felt understanding of how fire worked in such settings, much as we replaced the hearth with a propane heater and chandeliers with fluorescent lighting. Like the fusain-laden coal fed into power plants, fire's past was burned without our understanding what it contained. The fire paradox also has its intellectual dimension: the age of Enlightenment became a dark age for fast combustion.

2   *The Pleistocene*

On July 24, 1837, at Neuchâtel, Louis Agassiz, the young president of the Swiss Society of Natural Sciences, startled his audience by announcing that all around them was evidence of what he termed an *Eiszeit,* an ice age. His particular observations were not new. Other naturalists had recorded evidence of deep scratches on tough granite, erratic boulders far from their place of origin, and berms of rock and soil with no current mechanism to emplace them, and even local hunters and townsfolk explained such oddities by reference to glaciers now distant from these signs of former activity. In recent centuries, too, the ice had been moving down the valleys; there was no reason to think that it had not done so in the past and perhaps gone much farther. Agassiz consolidated those observations, borrowed a term from a colleague, and presented a theory that first shocked the young science—geology had only acquired a name 54 years earlier and had barely begun to appreciate the pieces and scope of its domain, what would become a millionfold increase in the known age of the Earth. In 1840 Agassiz elaborated on the *Eiszeit* with a book, *Studies on Glaciers.* Like his expanded publication, his vision of where the ice had spread had enlarged hugely. It had left alpine valleys to cover vast swaths of continents. It was, in a way, the Noachian deluge turned to ice.[1]

Predecessors like Jean de Charpentier and Ignace Venetz were alarmed at Agassiz's bold extrapolations; geology's elders, like Leopold von Buch, Jean Baptiste Élie de Beaumont, and even Alexander von Humboldt, dismissed it. Agassiz's Romantic vision of a frozen flood, however, caught the public's fancy. A younger generation turned his pronouncement into a research program and eventually an organizing principle for the recent epochs of Earth history. The Ice Age or, as further research revealed, a succession of glaciations had reshaped the Earth into a cryogeography. What wasn't frozen or under ice felt the effects of ice through water, wind, temperature, sea levels, faunal extinctions, and scores of other knock-on effects. Sea ice plated the polar oceans, land ice covered much of the Northern Hemisphere, and there was ice in soil and air and, it seemed, in the future. As late as the 1980s the consensus assumption was that humanity had been living in a halcyon age between glacial epochs, a skipped beat in the rhythm of planetary ice. The ice would return. It was a simple, irrevocable matter of math and geophysics. The Pleistocene laid down a template for what would follow.

It did, but in a way that passed ice through a looking glass and inverted it into fire. The debate over ice, hominins, and the Pleistocene would foreshadow the current conversation about fire, humans, and the epoch they have shaped. By the onset of the twenty-first century, the equivalent of Agassizes were organizing the scattered evidence of fire's coming pervasiveness and proclaiming the coming of the Fire Age. The disruptions of fire were proving as vast as those of ice, and the historical dimensions of a fire age, as controversial.

How the Pleistocene Got Its Ice

Planet Earth has swung into and out of hot and cold eras, greenhouse and icehouse worlds, before; ice ages have appeared not unlike the rhythm of mass extinctions. The worst, dubbed Snowball Earth, nearly plated over the planet in Precambrian times, perhaps several times prior to 650 MYA. Others followed—in the Ordovician, the Carboniferous, the Permian. The most recent icehouse world spans the Cenozoic but found its greatest expression in the Quaternary, the past 2.6 million years. It climaxes with a series of global glaciations that begin with—define the duration of—the Pleistocene.

The Pleistocene ice age wove together many geographic and climatic threads. How amenable the Earth became to ice depended on the amount and intensity of sunlight it received, how that warmth was distributed around the globe, and how it interacted with terrain. A uniform Earth would react one way. An Earth parsed into odd-sized seas and continents, with ocean currents, both deep (cold) and surface (warm), with plains and newly raised mountain ranges, all subject to changes, would react another way or, rather, many ways that themselves would change. The Drake Passage opened, separating South America from Antarctica. The Isthmus of Panama sealed off the Pacific from the Atlantic. The Himalayas rose, creating the Asian monsoon and stimulating erosion that buried carbon. Vulcanism spewed out carbon dioxide, a long-term agent of warming, and sulfate aerosols, a short-term agent of cooling. But mostly it appears that atmospheric carbon dioxide declined.

The icehouse world was preceded by a greenhouse world, one of the most intense recorded. The Eocene, 56 to 34 MYA, began with a thermal maximum, 5°C–8°C (41°F–46°F) above today's norm, and then slowly slid into a great chill. A global cooling set in during the Pliocene, roughly 3 MYA. The combined effects magnified the impact of Milankovitch cycles that characterize how the Earth wobbles, tilts, and stretches its orbit. Three rhythms are particularly noteworthy. The wobble refers to the precession of the equinox about Earth's axis of rotation, like slow gyres of a spinning top, that completes its cycle every 22,000 years. The tilt refers to the obliquity, or shifts in the orientation, of that axis relative to the sun; this cycles every 40,000 years. The stretch refers to the eccentricity of the Earth's elliptical orbit about the sun, a movement that beats to 100,000 years. These rhythms compound one with another to affect the radiant heat received by the Earth. The Milankovitch rhythms became a pacemaker of shorter-term cooling and warming. Cooling turned to ice; ice expanded into an ice age. About 2.58 MYA the first major glaciation arrived. It marks the onset of the Pleistocene.

How many glaciations occurred? It depends. Four great outbreaks, with lots of smaller bumps and bulges within them, are documented for the northern continents. A difficulty is that each surge can wipe out the evidence of the previous one as the new ice erases, scratches, and etches earlier records, relocates errant boulders, and smooths out moraines and potholes. The more extensive ice can obliterate the emblems of the less extensive. This process does not apply, however, to fossils deposited in the world sea. The record of oceanic sediments, based on oxygen isotope contents in foraminifera, captures some of the lost nuance and suggests perhaps forty or fifty glaciations, with a consensus of around forty-nine.[2]

The details matter less than the fact that, of the Pleistocene, some 80 percent was glacial, and of the past 900,000 years, 90 percent. There were internal shifts aplenty, including one that seems to favor the 100,000-year over the 40,000-year cycle, and

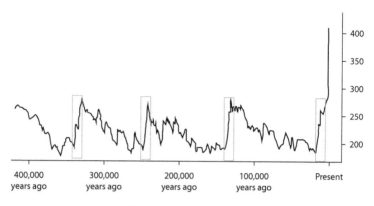

FIGURE 1. Glacials and interglacials. The pattern over the past 400,000 years as measured by CO_2 levels in parts per million and recorded in Antarctic ice, with the most recent levels supplied by the Mauna Loa Observatory. Rectangular boxes indicate interglacials. Data source: NASA, https://climate.nasa .gov/climate_resources/24/graphic-the-relentless-rise-of-carbon-dioxide.

of course local conditions modified whatever global movements were afoot. The cryosphere that appears so singular when viewed from afar looks messy when viewed up close. The interglacial phases were warm, sudden, and brief. The last glacial maximum crested about 21,000 years ago. The current interglacial began to dominate global climate about 11,000 years ago. It has displayed its own blips—heat waves like the medieval climatic maximum, or Medieval Warm Period (950–1300 CE), and cold snaps and the Little Ice Age (1550–1850) in which global temperatures dropped 2°C (3.6°F). The current interglacial has, by historical standards, lasted longer than previous episodes; has, until recent centuries, proved more stable; and has, beginning in contemporary times, been noticeably warmer and wetter.[3]

. . .

The Pleistocene remade the Earth's surface and rewired its dynamics. Ice was the most prominent feature, a defining presence, appropriately lending its name to the epoch, and it brought lots of secondary effects that encouraged a world still more favorable to ice. Ice became both cause and consequence. Its cryogeography dominated the Earth.

The ice covered one continent (Antarctica), spread across large swaths of two others (Eurasia and North America), and smothered one microcontinent (Greenland)—the Northern Hemisphere had much more ice because its land base was vaster and closer to the pole. Ice blanketed the two polar oceans, the Arctic and the Antarctic. It crusted major mountain ranges everywhere from the Andes to the Himalayas, claimed high valleys for cirques, and spilled down slopes in the form of glaciers. Even Africa's peaks had ice; Kilimanjaro boasted an ice cap of 15 square miles. Ice was in the air, the water, the soil. Surface ice had its counterpart in subsurface ice in the form of permafrost. The frozen ground fashioned patterns, an echo of the shapes that aboveground ice assumed. Rock glaciers, with an interior core of ice, shuffled down slopes.

Through wind, runoff, and uneven freezing and melting, the glacial realm extended its reach into periglacial landscapes. Melting soils led to debris flows, uneven thawing led to pocked thermokarst terrains in permafrost, hummocks rose into domes, ice wedges fractured surfaces, and frost-thaw cycles sculpted surfaces into polygons both tiny and immense. Glacial melt spawned streams. Ice sheets and glaciers birthed winds that could race down slopes with terrifying velocities. Glacial ice carried boulders (erratics), bulldozed berms (moraines), deposited hills (drumlins), and carved depressions (eskers and kettles). Glacial meltwater enhanced or redirected rivers. Glacial wind and water transported erosional debris ahead of the ice—deltas of sand called outwash plains, vast fields of silt known as loess. Ahead of mounded ice, lakes could form, ice could dam streams, and ruptures could occur that led to landscape-sculpting floods; and the pattern could repeat many times.

The ice had planetary effects. Its sheer weight could depress the solid earth. Even today solid-earth Antarctica is some 1,000 meters lower, on average, than it would be if all the ice were removed. Similarly, after 10,000 years Hudson Bay is still rising about 2 centimeters per year; the Baltic Sea, 11 millimeters per year. The Big Chill took continental-scale volumes of flowing water from the hydrosphere and froze it into solid-state ice on land. Antarctica still holds an estimated 60 percent of the Earth's fresh water. The magnitude of the freeze was enough to lower global sea level; at least four times, the ice surges removed enough water to drop sea level some 100 meters. The lowered seas exposed land—not just shorelines, but continental shelves, such that Britain was joined to France, Australia to New Guinea, and Siberia to Alaska. Pleistocene Australia was 25 percent larger than it is today. North America expanded to include nearly all of its continental shelves.

And the ice affected weather. A cooling climate had helped create the ice, but once it plated over northern seas and mounded atop so much of the Northern Hemisphere, the ice influenced ocean currents and atmospheric movements. It rearranged the patterns by which hot and cold air were exchanged, and those of warm and cold waters. The polar boundary that defines so much of weather in the Northern Hemisphere was broken and relocated. The effect was to reinforce the conditions that favored ice.

That weather was not limited to areas already with ice. In many places the shift meant cooler and wetter weather, which flooded depressions with lakes—most of the Great Basin in the western

United States was underwater (the Great Salt Lake is a tiny relic). Similar pluvial lakes extended along the mountains of central Asia (the Aral and Caspian Seas are the residuals). Lakes Makgadikgadi and Chad formed in Africa (a remnant of Chad remains). Elsewhere the changes meant cooler and drier conditions that led to desert, tundra, and steppe. The ice was not merely a product of the Pleistocene condition that reflected only itself: it projected its presence outward. Of course, not everywhere went glacial or periglacial, or trended toward extremes, but climatic sanctuaries and biotic refugia shrank and migrated.

Even when the ice retreated, its reach was powerful. Melting ice filled newly scoured basins and turned them into lakes—think the Great Lakes, continuing northwestward along the fringe of the Canadian Shield. Temporary lakes were common, ponding behind moraines and especially ice dams that could fail catastrophically. The channeled scablands of central Washington resulted when glacial Lake Missoula repeatedly broke through ice dams (perhaps forty times) and flooded the Columbia River gorge, plucking out fractured basalt, excavating channels, and depositing heaving mounds of debris in its wake. Much of the Earth's landscape still bears the marks of the Pleistocene, a world of grander forces and gigantic features that bequeathed a smaller-scaled world behind.

Yet discernable patterns are recognizable. Ice moves glacially, glaciations come and go with rhythms measured in tens of millennia, and icehouse eras rise and fall on the order of eons. In the short term, Milankovitch cycles influence atmospheric cooling and heating through solar radiation. In an intermediate term, the biosphere affects the storage and release of carbon, both on land and in the oceans. In the long term, carbon is stored in stone, in some fossil biomass like coal, and mostly in carbonate rocks deposited in the ocean.

Carbon, especially carbon dioxide in the atmosphere, makes a good marker for each episode, and much of its Pleistocene chronicle is preserved in the ice sheets still extant. The relationship is clear: more carbon dioxide, less ice; less carbon dioxide, more ice. During the last glaciation, an estimated 700 billion tons of carbon dioxide disappeared from the atmosphere. Perhaps 500 billion had been stored in terrestrial vegetation, now lost to ice and periglacial conditions; another 200 billion tons probably vanished into the oceans. The cycling of carbon dioxide follows predictable rhythms and can serve as a proxy for glaciations. Graphing atmospheric carbon dioxide roughly records the magnitude of planetary ice.[4]

The last great ice age crested around 21,000 years ago. The ice left the Baltic basin only 7,000 years ago. Some of the proglacial lakes still survive, if diminished. Lands continue to rebound upward from the sudden release of their icy burden. Life is yet reclaiming landscapes once occupied, then abandoned as the ice spread. Greenland and Antarctica still boast ice sheets. Permafrost continues to underlie vast extents of boreal North America and Eurasia. In the past at least some ice would likely have persisted from one glacial epoch to the next. Today, however, at a time when it might have strengthened prior to another surge, the Pleistocene's signature ice seems to be accelerating its retreat. Its refugia are shrinking. Almost everywhere the relict ice is receding or passing into extinction along with the vanished species with which it was associated.

Megafauna, Mega-extinction

All this change—a planet thrashing between frost and thaw, not just once but repeatedly—made conditions difficult for terrestrial life. Habitats changed, habitats shrank and expanded, habitats moved. Forests became grasslands, grasslands became deserts; shorelines moved down and out, then sank; valleys filled into lakes and then drained into playas; boreal landscapes vanished into ice-scapes. As always, some species could adapt, some couldn't. The normal rate of extinction quickened. Unsurprisingly, over its 2.6-million-year duration, the Pleistocene held the last of the Earth's five great extinction events.

The geologic record favors the big and the hard—large animals with bones—over the small and the soft tissued. Even so, the ice ages seemed especially hard on megafauna. Genera and species expired, then reappeared in new avatars. Globally, of 714 large mammals known from the fossil record, 207, or 29 percent, went extinct. The process was not continual: some times and places experienced greater or lesser impact. Extinctions were less severe in large land masses and in places close to the evolutionary origins of genera. Extinction crested in the early Pleistocene for Africa, and in the late Pleistocene (and early Holocene) for Eurasia, North and South America, Australia, microcontinents like Madagascar, and islands. It was slowest in Africa and Eurasia, more sudden in the Americas and Australia, and virtually instantaneous for true islands. In Africa some 70 percent of large mammals survived the gauntlet. In North America the figure is closer to 25 percent, and for Australia, roughly 5 percent. On microcontinents like Madagascar and New Zealand, none survived.[5]

This pattern held for the hominins no less than for other mammals. The clade radiated into many species, then shrank, then reradiated. The longest lived was *Homo erectus,* whose evolutionary tenure at roughly 2 million years encompasses most of the Pleistocene. The species may have survived to as late as 50,000 years ago on island refugia like Sumatra. At the onset of the last great glaciation, there were apparently as many hominins as there were elephants. Just as there were mammoths, mastodons, ancestral African and Asian elephants, dwarf elephants on most Mediterranean islands, at least a dozen in all, so, too, there were habilines, erectines, and erectine spin-offs like *H. heidelbergensis*. There were Denisovans, *H. floresienses*, Neanderthals, and for at least the past 200,000 years, sapiens. All flourished during the interglacial prior to the last glacial, which reached its maximum around 21,000 years ago. Only the sapiens have survived.

Simple island biogeography accounts for much of the species variance, but not all, and not for the tempo of more recent events. Evolution had accommodated the pulsing of the ice in a kind of running biotic gloss. New species appeared, new species died out. But the most recent epoch broke that pattern. Time and again, especially during the Holocene, humans drove creatures over an evolutionary cliff. Just when and how the new pattern of hominin-catalyzed extinction replaced the old one is a subject of continual research and, unsurprisingly, of controversy.

It was an age of gigantism. The Pleistocene boasted ice sheets the size of continents, oceans chilled into frozen gyres, immense lakes, and vast deserts, all populated by big creatures. Everything, it seems, that exists now existed then in a larger form. The Norse legends of Jotuns and giants in a land of ice had their echoes in reality. North America had ground sloths as big as Asian elephants. Antarctica boasted penguins the size of Shetland ponies. New Zealand hosted moas twice as tall as people. Australia had carnivorous goannas 7 meters long. Even hominins were bigger—the Neanderthals were the largest of known hominins and had a more capacious cranial cavity for a bigger brain. The era left a legacy landscape larger than the processes it later contained. It's as though today's smaller Earth dressed in clothes that no longer fit.

The rhythms of ice and thaw, extinction and emergence, persisted throughout the Pleistocene. Until they didn't. Many genera didn't rebound, and megafauna, and then hosts of species, continued to decline, save for those protected by humans. The ice didn't return, then it continued to melt away. The anticipated storage of carbon stopped, then reversed. The Earth warmed, and then it continued warming. The critical difference that separates this interglacial from earlier ones is the singular presence of sapiens.

How the Pleistocene Got Its Tale

It had been an unstable epoch. It had filled and emptied repeatedly, as ice advanced and receded, lakes rose and fell, and biotic migrations radiated outward and retreated back. Its defining physical property was the tireless compounding of solar radiation with land, sea, and air. But the epoch also proved intellectually unstable. Its temporal domain changed at both ends, unsettling a narrative. Even its name has its history.

Time was shallow before it became deep. Until the nineteenth century, scholars believed the Earth had existed for some 6,000 years, based on chronologies extracted from the Bible (even Isaac Newton did the numbers). In 1756 Johann Lehmann identified two broad categories of rock: an older (Primary) and a younger (Secondary), overlain by recent soils. Four years later Giovanni Arduino proposed to enlarge the newer layers into a Tertiary era. Comte de Buffon boldly proposed an order-of-magnitude increase in the Earth's age, to 60,000–75,000 years. Five years later, in 1783, the term *geology* was coined, and the new discipline took the age of the Earth as a defining question and the organization of deep time as its theme.[6]

The breakthrough came when Charles Lyell created a more detailed geochronology in his three-volume *Principles of Geology* (1830–33). Lyell subdivided the Tertiary into four phases. He called the oldest epoch the Eocene, deriving the term from the Greek *eos* ("dawn" or "earliest") and *kainos* ("recent"). He called the next oldest the Miocene, from the Greek *meion* ("less"). Then came the Pliocene, from *pleion* ("more"), divided into Older and Newer. Beyond that, leading to the present, was the Post-Tertiary. (Jules Desnoyers had proposed, a few years earlier, a separation of the more contemporary times into a Quaternary era, but Lyell disliked the term.)

It was the Post-Tertiary that struggled to find its footing. Even the name was conflated, with the Latin *post* mixed with the Greek *kainos*. Lyell divided it into two phases: the Post-Pliocene and the Recent, and of course *Post-Pliocene* immediately got confused with *Post-Tertiary*. He urged that *Recent* be used to describe the period of "known human presence," which at the time barely extended beyond the chronicles of the Bible. Within 30 years Lyell himself would extend human history into Earth history with the *Geological Evidences of the Antiquity of Man*.

Barely had the ink dried on the *Principles of Geology* than Agassiz issued his dramatic annunciation of an Ice Age. In 1839 Lyell replaced *Newer Pliocene* with *Pleistocene,* advancing the Greek from the comparative to the superlative (*pleistos* and *kainos,* "most recent"). He quickly regretted the decision and withdrew it, not wishing to distinguish too closely the near from the now. But Edward Forbes reinstated the term in 1846 as a synonym for the *Ice Age,* and it has been used ever since, despite Lyell's reluctance and stubborn continuance of his original term in later texts.

Similarly, Lyell's use of *Recent* was challenged in 1867 by Paul Gervais's *Holocene* (from the Greek *holos,* meaning "whole"; why he chose the term is unclear). Together, Pleistocene and Holocene made up a composite era, the Quaternary, which fully replaced Post-Tertiary. In 1873 Lyell accepted Forbes's redefinition; 2 years later he died, and resistance ended. In 1885 *Holocene* was formally submitted to the International Geological Congress and became the preferred working term.

Still, the Quaternary—Pleistocene and Holocene—remained the problem child of the geologic timescale. While other eras were defined by strictly geologic criteria and mass extinctions, the Quaternary was organized around ice ages and the "tenancy of man." The Holocene embraced the age of humanity, as the Pleistocene did the age of ice. But this left matters unsettled and ambiguous because glaciation was not simultaneously global (many areas had lakes and deserts instead of ice), the tenancy of humanity kept being pushed back to the origins of that glaciation epoch, and there was no clear-cut inflection from Pleistocene to Holocene, since there was no evidence that the cycling of glacials and interglacials had suddenly ceased around 10,000 years ago. Instead, the epoch became an anomaly, defined by climatic boundaries recorded through ice and the relict geomorphic "fossils" of glaciation rather than by extinctions recorded in fossils of organic evolution. So even as the term *Pleistocene* became fixed, the Pleistocene epoch's borders remained elastic, its duration unknown, and its age only relative.

Steadily, its dominion expanded. In 1863, 4 years after Darwin published *On the Origin of Species,* Lyell estimated the duration of the Pleistocene at 800,000 years. In 1900 W. J. Sollas put it as 400,000 years. In 1909, based on Alpine terracing (wrongly, as it happens), Albrecht Penck and Eduard Brückner dated the epoch's start at 650,000 years. As a rule of thumb, working geologists bracketed the epoch as "the last million years." In 1948 an International Geological Congress combined the glacial with the paleontological indices and identified the lower boundary of the Pleistocene with the advent of cold-loving species in the Mediterranean Basin, some 1.64 MYA. This event was recorded in the Vrica stratum in Calabria, southern Italy, which marked "the horizon of the first indication of climatic deterioration in the Italian Neogene succession." This conclusion was affirmed by the International Union for Quaternary Research at its seventh congress in 1965 and again in 1983. Granted that organisms had begun colonizing the continents around 430 MYA, the Pleistocene constituted less than one-half of a percent of the estimated duration of terrestrial life.[7]

For 176 years after Charles Lyell had identified and named the Pliocene, and 170 after he had coined the term *Pleistocene* to replace Newer Pliocene and Post-Pliocene, the marcher lords of geology had fought over where to draw the boundary between the two epochs, and for all that time the Pleistocene had resisted the demand that it be like all the other realms of geologic time. It was proudly, stubbornly anomalous. It demanded, as it were, a separate creation based on its own peculiar traits as it defined them. It was the epoch of ice and human origins.

In the early twenty-first century the International Commission on Stratigraphy tried to abolish *Quaternary* as an obsolete term for an idiosyncratic period, treating it as a miscategorized entity, rather as astronomers did Pluto. The Quaternary community rose in protest and actually succeeded in enlarging their domain. Instead of disappearing, the Quaternary expanded, in what some observers sourly called a "land grab," as the Pleistocene encroached into the

Pliocene to the amount of 800,000 years. This put its official start at 2.58 MYA and left the Pleistocene 260 times longer than the Holocene. The Pleistocene had proved as unstable as its climatic rhythms, its spasms of extinction, and the self-regard of its signature species. In 2009, bowing to the Quaternarists, the International Commission on Stratigraphy approved the change by a vote.[8]

Among the objections of geologists not working in the Quaternary was the observation that the criteria used to end the Pleistocene were different from those used to begin it. The narrative starts with one theme, climate, and concludes with another, humanity. The processes that underwrote an ice age did not cease 10,000 years ago, nor did the hominins expire. The Holocene seemed an arbitrary creation. The creature that was supposedly using objective evidence to parse deep time was talking about itself. The tale of the Pleistocene looked like something told by an unreliable narrator.

This of course did not end the paradox. The Pleistocene is some 260 times as long as the Holocene. Plotted according to strictly geologic criteria, the Holocene is just another interglacial, and probably one primed to implode back into ice. What justifies its autonomy is exactly what Lyell noted: its association with humanity. Humanity, however, continues to nudge its origin story further back, to where the hominins now span the Pleistocene. Even the sapiens are at least 200,000 years old. The Holocene has become the epoch during which the sapiens spread across and even beyond the planet. Only humanity's vanity warrants splitting off the last 10,000-plus years as an autonomous epoch in the geologic record. But then humans are the ones who get to choose and name. Humans, it seems, have unsettled the identification of geologic periods as their presence has remade them physically.

Keeper of the Flame

The debate over ice, hominins, and the Pleistocene foreshadowed the current conversation about fire, humans, and the epoch they have shaped. The disruptions of the emerging Fire Age are as vast as those of the Ice Age, and their historical dimensions as controversial. By the onset of the twenty-first century, observers were scrounging for the right word to describe the cumulative range of anthropogenic fire's impacts and its temporal bounds.

The sapiens were a late-stage hominin, the last we know of and the only one who survived the fifth extinction. Many hominin species had overlapped, then they didn't. There was some interbreeding between sapiens, Neanderthals, and Denisovans. There was probably some conflict. But the sapiens survived where the others didn't. There is some evidence and argument that the demographic bottleneck was narrow, but the sapiens passed through it and spread throughout parts of Africa, the Middle East, and Europe. Then they continued along paths already well worn by erectines and others. By 50,000 years ago they passed into New Guinea and Australia. The great surge that took them beyond the prior range of hominins had to wait for the last glacial maximum to recede and open up the Americas. Islands followed; hominins continued colonizing uninhabited isles throughout the nineteenth century. They established permanent bases in Antarctica only in the latter half of the twentieth. Soon after, on plumes of fire, their voyaging left Earth.

They were a greater disruptor than the others. In too many cases is the synchronicity of megafaunal extinction and the arrival of the sapiens too close to ignore. The geography of extinction seems to track the spoor of the sapiens. The record is clear for historic and prehistoric times: the last extant humans wiped out countless species on first-contact islands. The difficulty is to project that documented practice into the more distant past and across vaster landscapes. The sapients were omnivores, which also means they were predators; they had social organization leveraged by language; they had novel hunting technologies; and they had fire. Together, that was enough to shake the branches of the tree of life and let other species fall. The sapients began reshaping biomes.

But a similar detour characterizes the course of carbon dioxide and methane. Instead of steadily eroding from the atmosphere, the gases began to increase. The warming that was underway—what seemed only the most recent in a relentless cadence of climatic comings and goings that characterized the Pleistocene—left its beaten track and wandered into planetary wilds. As with extinctions, the most plausible cause is the propagation of the sapients. The biotic radiation of the species challenged, in complexity, the solar radiation of the Milankovitch cycles. The geophysical warming that inaugurated the Holocene took on a life of its own no longer responsive to the old cycles.

The sapients, like all the hominins, were an ice age creation. But they were also a fire creature and, in time, the only one on Earth. Their firestick became an Archimedean lever that, given a favorable fulcrum, they used to move the planet. With the last interglacial, a fire-wielding creature met a fire-receptive environment. For millennia after millennia, aspect after aspect of Earth— from the atmosphere to geochemical cycles to the biosphere—was nudged, then hefted, into a new arrangement. Each ratchet seemed to magnify the sapients' fire power. Fire begat fire. The Pleistocene segued into a Pyrocene.

It was easier to live with fire than with ice. Ice is exclusive: it buries rock, destroys life, denies anything but itself. It results from the pure mechanics of water and cooling. It can occur without a particle of life present. It's tangible. It persists, with long lag times, even after its originating conditions have changed. But fire is a reaction, it's fleeting, and it depends utterly on life, which supplies both the oxygen and the fuel it demands. Even a single fire can vary as it burns through different terrains, weather systems, and vegetation. It can be manipulated in ways ice cannot.

Against an approaching ice age, humans could only leave or adapt at the margins. Amidst an approaching fire age, they could seize some of the critical levers and move the machinery. Terrestrial life had coevolved with fire, tinkering with adaptations and remaking the character of combustion. The hominins had an additional advantage in that they could kindle fire. The sapiens went further. They could kindle a geologic epoch.

3 *Fire Creature*

Living Landscapes

His uncle had raised him, and when his uncle died, Thomas Livingstone Mitchell, then 16, enlisted in the British Army and fought through the Peninsular campaign. At 19 he was a lieutenant in the Ninety-Fifth Rifles but occasionally seconded to the quartermaster-general's service because of his skill as a draftsman. When the war ended, he returned to survey and document the battlefields in what became the classic Wyld's *Atlas Containing the Plans of the Principle Battles, Sieges, and Affairs* of the Peninsular War. Those same talents led him to be appointed surveyor-general for New South Wales in 1827.

Mitchell sketched and mapped first around Sydney. Then he began a series of four expeditions, each taking him farther afield and establishing him as one of the premier explorers of interior Australia. He wrote a narrative of his first three expeditions in one book and of his fourth, to Queensland, in a second. They proved rich sources of detail about the land, its natural history, and its indigenes. While irascible and testy, particularly with those in authority (he was the last known person to engage in a duel in Australia), Mitchell recorded Aboriginal Australians with a sympathy and astuteness not typical for his time. Inevitably, his observations included fire.

In the basin of the Darling River, he noted the "annual conflagration of the countryside." Elsewhere he recorded "the extensive burning of the natives, a work of considerable labor. . . . We perceived that much pains had been taken by the natives to spread the fire, from its burning in separate places." The party rode 5 miles through "this fire and smoke," the latter of which "added much sublimity to the scenery." Most places were burned, and Mitchell was hardly alone in recording them. Because British contact with the Aboriginal Australians came relatively late, the Enlightenment had already radiated widely, and explorers from Europe had training in natural science or had naturalists accompany expeditions. Landscape was something they sought to record in ways unlike those of Iberian explorers, whose chroniclers were conquistadors and clergy.[1]

But Mitchell had a richer understanding of landscape fire, almost unique among his generation. In a famous passage he elaborated on how it interacted with flora, fauna, and people:

> Fire, grass, and kangaroos, and human inhabitants, seem all dependent on each other for existence in Australia; for any one of these being wanting, the other could no longer continue. Fire is necessary to burn the grass, and form those open forests, in which we find the large forest-kangaroo; the native applies that fire to the grass at certain seasons, in order that a young green crop may subsequently spring up, and so attract and enable him to kill or take the kangaroo with nets. In summer, the burning of long grass also discloses vermin, birds' nests, etc., on which the females and children, who chiefly burn the grass, feed. But for this simple process,

the Australia woods had probably contained as thick a jungle as those of New Zealand or America, instead of the open forests in which the white men now find grass for their cattle, to the exclusion of the kangaroo.[2]

Mitchell then noted the melancholy consequences that followed the "omission of the annual periodical burning by natives. . . . The open forest lands nearest to Sydney" yielded to "thick forests of young trees, where, formerly, a man might gallop without impediment, and see whole miles before him. . . . Kangaroos are no longer to be seen there; the grass is choked by underwood; neither are there natives to burn the grass."[3]

Talented, cranky, a seasoned traveler, a shrewd observer, T. L. Mitchell sensed how fire functioned in Australia, that it was neither vandalism nor a simple tool like a boomerang or a digging stick, that it wove through the tapestry of the living landscape, that its removal could be profoundly disruptive, that fire and people shared an intimate alliance that propagated through everything else. Those conclusions are not limited to Australia.

Earth is not alone as an ice planet. Mars has ice, as does Earth's moon, and as do many moons of the outer planets, some of which (like Enceladus and Triton) are wholly ice worlds. But thanks to life, only Earth is a fire planet. When life evolved a creature to manipulate fire, their interaction proved not additive but exponential. A second-fire appeared to challenge the first-fire of nature.

.   .   .

Hominins did not invent fire. They captured it from their surroundings.

Fire was all around them in the savannas and woodlands from which they emerged. They scavenged through the ashes, foraged amid the rich regrowth, and occasionally fled bursts of flame. At some point they picked up burning branches. They saw they could kindle unburned patches of grass. They learned that, instead of scrounging for burned meat, they could burn it on their own. They could cook. If they fed the flame, they had light and heat. Around the flames, they could gather—those who share a fireside was for a long time the definition of a family. They merged fires to signify marriage or treaty. To keep the home fires burning was an emblem of endurance. Around the flames they recounted the day's experiences, crafted the stories by which culture was transmitted, and danced and sang and enacted the rituals and ceremonies that defined who they were. Over and again, origin myths speak to a frail and threatened species that became powerful only when it acquired fire. Fire was presence, tool, and companion.

But above all, fire was power. It was more critical for people than for the rest of nature. Most marine life did not need fast fire, terrestrial life had to accommodate it and could actively promote it, but human life could not exist without it. Some species could thrive in fire-immune environments; hominins could not. Fire was not simply something people adapted to or exploited selectively for relative advantage. It was essential to their existence. Thanks to cooking, it entered their genome. Before anthropogenic fire changed the world, it changed hominins.

Second-Fire

The hominin acquisition of fire marks a phase change, not only in fire history, but in Earth's evolution. In some respects, hominins and their fires simply folded into the endless, complex processes that have shaped how nature's fires have appeared on landscapes. Together, they offered a new source of ignition and additional means to rearrange the vegetation that served as fuel. The ignition could be more pervasive and regular, and the fuel fussing could be more varied, but they operated within the same constraints of nature's fire economy. So if humans had their hands on a lever of immense power, they still had to operate within the existing machinery. They could interact, they could massage, but they could not dictate.

As with other species exposed to fire, they could change the character of fire even as fire changed them. Their relationship to fire went beyond prey-predator models or simple tool use to project their will. It came to resemble a mutual assistance pact in which each party—fire and hominins—empowered the other. With human help, fire expanded into realms it could not have claimed on its own; with fire, humanity did the same. They journeyed to places otherwise inaccessible, not only to themselves but to life. They went to the Atacama Desert and the Greenland ice sheet; they sailed below the Arctic pack ice and circled the moon.

The upshot has been an acceleration in fire's presence—a minimal increment at first, profoundly modulated by the flickering cycles of frost and thaw; then a more modest power, as the hominin hand and mind joined fire to capture fire-prone landscapes as they had captured fire; and finally, with the sapiens, an unstoppable force. By cooking food, the hominins got small guts and big heads. By cooking landscapes, they went to the top of the food chain. Ultimately, by cooking the Earth, the sapiens became a planetary power.

There is an old philosophical discourse about whether humanity is fully a part of nature or separate from it. How much of us is nature and how much nurture? The entanglement is particularly complex for fire. The saga of humans and fire begins as all nature and ends as mostly nurture. It begins as anthropogenic fire competes with natural fire, subject to the same restraints and propulsions, feeding on the same fuels. It ends in recent centuries with all parts of fire's essence separated, bolstered, and reassembled. Anthropogenic fire became a novelty so radical that it challenged the way fire operates on a planetary scale. Eventually humans reshaped every aspect of fire's physical environment. They reorganized the geography of natural fire. They fashioned a nurtured doppelganger to nature's fire.

So profound is that inflection point at which a fire-wielding species met a fire-receptive epoch that it warrants a separate consideration as perhaps the grandest change in fire on Earth since plants colonized the continents. Natural fire—the first-fire—had a challenger, one that deserves a separate title. Call it second-fire, for it operated within what Cicero famously called a "second nature" that human artifice had fashioned from first-nature.

Fire's Crucible

The critical technology, the foundational pyrotechnology, was cooking. The controlled heating of foods is a process of predigestion. It softens tough fibers, chemically enriches carbohydrates, renders edible some potential foodstuffs otherwise inaccessible, detoxifies tubers like cassava, and purges meats like pork and bear of parasites like trichinosis. Research seems to demonstrate that people cannot survive—cannot reproduce and thrive—on raw foods alone. For early hominins, cooking furnished a critical increment, a nutritional tipping point, that allowed for smaller guts and less robust mandibles, larger crania, and the energy necessary to power the bigger brain that occupied that space.[4]

Cooking became, in turn, the model for most pyrotechnologies. Over time people learned to cook sand, ores, clay, mud, limestone, wood, and oils to create glass, metal, pottery, bricks, cement, tars and turpentine, and assorted potions and aromas. They used fire to make other tools, from fire-hardened spears to plows to cannons. Few technologies did not involve fire at some point in their chain of creation. Fire was interactive: fire as a tool could make other tools.

Watching fire make their world—and in primeval times, fire loomed larger because it had so few technological rivals, was not so sublimated within complex manufacturing—it was easy for people to imagine the natural world likewise made by fire. In the dawn centuries of Western civilization, fire was not only something that needed explanation, but something by which to explain. It could make understanding as it made food and metals.

For the spiritual world, fire was a foundational god. Yahweh manifested himself to Moses through a burning bush. Of the twelve gods on Mount Olympus, two were fire deities, Vesta and Vulcan. Agni was the god of fire in India. Dios Viejo, the "old god," was an originating deity of pre-Columbian Mexico. Candles and eternal fires on the altar accompanied rituals. Sacrifices were typically burned to send their smoke to the heavens. New fires were kindled during crises and at prescribed moments of cosmic significance like the Aztec New Fire rite that came every 52 years and gave the passing world a new sun. The ceremonies testified to the power of fire to improve the good and destroy the bad, to shatter and renew whole worlds. Even when the underlying reasons or the originating historical circumstances no longer survived, the fire rites often did.

The same held for the secular world. For Heracleitus, fire was a universal emblem of and means for change: all things were an exchange for fire, and fire for all things. For Empedocles it was one of four basic elements. For Aristotle it was the model system for change; no system for explaining the world could work until it accommodated fire, because all change, and the world was endlessly changing, was based on fire. The alchemist's lab was the hearth outfitted for cooking materials in order to coax the transmutation that fire wrought in so many substances. One of the founders of the Royal Society of London, Robert Hooke, reported studies in which charcoal was examined under a primitive microscope in *Micrographia* in 1666, and then as surveyor of London he mapped the Great Fire of that year. Even into the early Enlightenment, "fire" was everywhere: in the heavens as sun, stars, and comets; within the Earth as magma; on the surface as the endless flames of nature and humanity.

Beyond status as a tool or a concept, however, fire was a relationship. It may well be humanity's first domestication. Unlike an axe, scraper, or spear, it could not be put on a shelf and ignored until it was next needed. Once kindled, it had to be tended. It had to be fed, housed, trained, watched. Someone had to care for it, continually. (The English terms used are the same as those used for rearing children.) If it sparked a social order for gathering to cook and converse, so it reorganized a group to provide for its needs; furnishing fuelwood alone could consume hours daily. If not continually maintained, it still required all the ingredients present so it could be made at will. The Iceman who was recovered from a glacier in the Ötztal Alps had few possessions on him, but he did have flint and tinder. Still, the preference was to keep fire constantly alight. A crew of French sailors and Malagasy slaves shipwrecked in 1761 on Île de Sable (Tromelin Island) east of Madagascar managed to maintain fires for cooking and signaling for 15 years before eight survivors were found.[5]

This demanded a place for the fire, one sheltered from rain and other elements that might extinguish it. The *domus* was as much for fire as for humans. The first species to be domesticated was not a species at all but a peculiar process, a reaction that was not alive but came from the living world and so had some of its properties. It was a model for what followed. To domesticate a species was to bring it into the domus of a shared fire.

. . .

The hearth fire did not stay in the hearth. It returned, transformed, to the landscapes from which it had come. Surely, much of this occurred by accident (as it still does today). But for creatures attracted to burned landscapes and the foodstuffs that flourished in the ash, the notion would take hold that they could create such conditions themselves and not rely on nature's whimsy. They saw how many of the animals they hunted were drawn to the fresh forage that sprouted after a burn; if they burned on their own terms, they could move those creatures around a landscape in ways favorable to themselves. They could tweak, not simply adapt. They could hack into and manipulate the operating system.

As a thought experiment, consider bison, which graze preferentially on the most recently burned sites; almost none appear on unburned patches 2 years after a fire. Suppose that the bison could start fires, and consider what power that would grant them over their feeding, its capacity and timing, as well as over the larger ecosystem that their fires rambled across. Yet that is exactly the position humans found themselves in. In ways unimaginable for other species, they could shape not only their immediate habitat, but the larger environment in which that niche occurred. Fire's power to transmute and propagate cast its influence widely. Ultimately, the torch—the firestick—could act as the ecological equivalent of a biotic lever by which to move landscapes.

Such a lever needed a fulcrum, and that resided in nature. What humans could manipulate with aboriginal fire economies—*aboriginal fire* here meaning "fire practices based on the control of ignition"—was the placing and timing of fire starts; and by their repeated burnings they could alter the character of the land and its amenability to being burned. Whether a fire snuffed out or spread depended on environmental conditions over which those who held a firestick had little say, other than to set the fire under conditions that encouraged a fire to burn as they wished. Yet the process was, if in a looser way, a means of extending domestication beyond the realm of the hearth. The torch bound campfire to country fire.

Aboriginal Fire: Early, Light, Often

Aboriginal economies show a remarkable consistency across environments. Similar patterns, both spatial and temporal, appear in boreal forest, tropical savanna, desert steppe, and semiarid and mixed woodlands; they just express themselves differently. Spatially, aboriginal fire appears as lines of fire and fields of fire. Temporally, it typically arrives before the natural fire season or during late-season curing as part of a harvest or hunt. It can drive game to where they can be more easily hunted over the winter; it can make gathering acorns and chestnuts easier. The fires come early and often, before the rains; or late, after dormancy and prior to snow.

Together, they inscribe a matrix for pyrogeography. Lines of fire track human movement across landscapes; some fires are set to improve that mobility, and some (like signal fires and abandoned campfires) are just by-products of a flaming companion that accompanies people everywhere. Fields of fire are patches people deliberately burn to improve hunting and foraging or to maintain fire-protected cordons around settlements. Places with natural fire have their pyric realms expanded and recoded.

Something similar happens across time. People return on some cycle, perhaps annual, perhaps decadal, adjusted to actual conditions. In naturally fire-prone landscapes they kindle their fires prior to lightning's appearance. They begin with small patches, then increase them in number and size as the land dries or cures, abutting one patch against another. By the onset of the rainy (and lightning) season, all the areas that people want burned, or shielded from wildfire, have been fired. Guidelines may be natural, such as burning when soil is moist but fuels dry; or they may follow experience as coded into rituals, songlines, or other ways of recording and prescribing social knowledge. Doing this year after year, century after century, millennium upon millennium, forces everything in that countryside to adapt. As fires become more regular, they become easier to control.

Fire integrates, fire interacts. People did more than burn, and whatever else they did, their fires also expressed. The firestick had to engage with digging stick, spear, and boomerang. Perhaps the most interesting interaction was with the abundant megafauna, because what was forage for herbivores was equally fuel for fire. If megafauna died out in fire-prone settings, their removal liberated more fine fuel for fire. The land became more fire prone. If they disappeared from fire-intolerant places, like shade forests amid climates that lacked wet-dry rhythms, it could mean the abolition of free-burning fire. As the sapients spread, megafauna vanished. In central Europe that meant a loss of fire and a shrinkage of habitat capable of supporting people. In most of Australia it meant a proliferation of fire and the promulgation of habitats better suited to aboriginal economies. In North America outcomes varied as climates, fire, and fuel jostled into accommodations after the ice had receded and left some places more fire prone and some less.

The firestick could be both precise and ambiguous; it could operate at scales both tiny and huge. It could burn cavities in tree trunks, which then (perhaps with the assistance of termites) could be hollowed out to create spaces attractive to nesting mammals, which could then be smoked out and hunted. It could burn berry patches and willow thickets to stimulate fruiting or new twigs suitable for baskets. It could bait snares with fresh regrowth or produce smoke to drive off mosquitoes and blackflies and attract deer or rhinos. Its light could "spotlight" prey and attract fish at night for easier spearing. A smoking firestick could stun bees out of their hives and make it easy to gather honey. Broadcast burning could operate at landscape scales, nudging herds of grazers into pastures for winter and summer or drawing them into mountain valleys where winter snows would block the passes and seal them into a suitable hunting ground. Anything that could burn, anything that could be gathered, stalked, or trapped—people found the ecological sweet spot where their fire could make them accessible. Combustibles became consumables.[6]

In a famous 1969 essay, Rhys Jones described the cumulative consequences of Indigenous burning in Australia as "firestick farming." So abundant, so frequent, so calculated were the fires that they added up to a kind of pyric horticulture. It was not agriculture as European agronomy understood farming, since it lacked plowed fields, fenced pastures, fixed landownership, crop rotation, and so on, but it made a cultural landscape out of the indigenous materials as much as Europe had with its biota. This was not a wilderness through which the native peoples passed like wombats or the wind. It was a fashioned landscape made with a simple but powerful technology ideally suited to Australia, fire.[7]

Still, it can seem fantastic. Australia was large and its indigenes few; stone and wood tools are feeble, and landscape fire is diffuse. Intuitively, we equate more fires with more people. But in aboriginal economies people can wander over vast regions as part of seasonal cycles of hunting and harvesting, spreading fires in ways unimaginable to European peasants legally tethered to fixed plots of land. They can start fires in abundance, if we count all fires, especially if they carry smoldering firesticks with them as they travel. Rhys Jones estimated that in the better-populated areas of Australia, an area of 30 square kilometers would have supported a band of roughly forty people. "Assuming that on average, three foraging parties of various types left camp per day, that each lit ten bushfires and this happened on only half of the days of the year, then within that area, no less than 5,000 separate bushfires would be lit each year"; he called this a "highly conservative estimate." Besides, fire as well as people can wander. It is not like a plow that must be pulled down every furrow, or an axe that must fell every tree; once kindled, a fire can spread, self-propagating, until rain, faltering wind, or stingy fuel ends it. More people can mean more fires but also more-intensive fire control. The ideal formula for landscape fires is for the firestick to go on walkabout.[8]

How do these fires interact with nature's? They can both compete and collude. At a given place and time, only one can exist:

what one burns the other cannot. Here they compete, and by deft burning, aboriginal fire practices can prevent wildfires where they might threaten settlements or hunting grounds. But the fires can also occur outside the natural cycle or in places (think Mediterranean climates) that are fire prone yet fire infrequent because they lack routine lightning. Here they collude, with humans adding the ignition that nature cannot regularly furnish. Over more and more places, second-fire began to supplement, and then replace, first-fire. A tamed country supplanted a wild land.

The power of such fires, which can seem no more consequential than a rain shower, has become apparent in recent times with the creation of nature preserves and protected forests, which promptly sought to exclude such fires as unnatural and damaging. In some locales, the landscape has shifted from one biome to another, with woods, say, replacing grasslands or an unbroken monoculture of trees filling in a patchy forest. In others, after a pause in burning, fires have returned, often savagely and outside historic bounds. Remove the firestick, and feral flames replace tame ones.

Agricultural Fire: Fire and Fallow

Yet aboriginal fires came with severe limits. They could only burn as their environment permitted. They could burn in some seasons and not in others, amid some biomes more than others, through wind and sun but not in fog and snow. People could tweak what existed to their advantage—and so sought out places amenable to fire. They could not bring fire to sites whose natural endowments made fire unlikely if not impossible. If they wanted fire, they would have to alter the fundamentals.

This, for fire history, is the meaning of agriculture. By slashing and drying woodlands and peat, by draining wetlands, by loosing livestock to trample brush and trim shrubs, people could create conditions that made fire possible, and then by introducing plants adapted to postburn conditions, they could grow food. They could notch up their fire power from firestick farming to a regimen of fire-fallow farming. People and fire could push against the old constraints. Agriculture could relocate outside floodplains; fire could expand into sites, and at times, it could not otherwise. Even today, most open flame in living landscapes occurs within an agricultural context shaped by stubble fires, fallow fires, slash-and-burn cultivation, pastoral burning, conifer plantations, and land clearing in rainforests and peatlands.

Burning did what fire ceremonies proclaimed for it: it promoted good and purged bad. It fumigated and fertilized—temporarily driving off the indigenous flora and fauna, creating conditions that allowed species from elsewhere to thrive in the ash. It helped of course if those flora and fauna came from fire-adapted biomes, which most did. The hearths (a revealing term) from which domesticated plants have come are all in regions that undergo regular wet-dry cycles and so know fire and its aftermath. The source regions for most domesticated fauna are likewise regions of routine wetting and drying, and often mountainous regions where pasture can be found by moving seasonally up and down slopes—more areas naturally adapted to fire. For both flora and fauna, people could supply ignition.

A fire-catalyzed agriculture is an exercise in applied fire ecology. It relies on the ecological jolt furnished by burning to renew the site. That demands suitable fuel, and in an agricultural system, that means growing those combustibles. This is the most plausible explanation for fallowing, a practice that mystified European agronomists, who deplored it as wasteful and denounced it as unfounded folklore. Worse, to their minds, the fallow was burned. But the fallow wasn't burned in order to get rid of it; it was grown in order to be burned. The fire was critical to the system, and it needed fuel.

There were many ways to expand the realm of fuel. Farmers could ringbark or fell woods, then leave them to dry. First-contact slashing was more onerous but could yield richer crops. In temperate and boreal woods, large trees were typically logged off, or killed but left standing so they would not clutter up the field. In tropical woods a few might be left to furnish some useful shade. What burned best were the fine fuels—the small trees, the shrubs, the surface litter. The point was to amass as much of these as possible and spread them across the ground to promote a hot, even burn. If the site lacked enough biomass, additional fuels could be brought to it. Needles and branches, dung and straw, even dried seaweed—a good fire needed ample fuel.

Landscapes lacking woods could supply other sources in the form of organic soils such as peat from either upland moor or wetland (this was a major ambition of the agricultural revolution in Europe that preceded the industrial revolution). Instead of being slashed, the site would be drained, with the depth of the water table determining the available fuel. A shallow plot could be burned as it was; deeper exposures could lead to paring and burning, in which the sod could be cut, stacked, dried, and then burned. Shrubs might be cultivated in rows in which their branches could be lopped off and left for burning—coppice for field fuels. In the Canary Islands, pine branches served the same purpose. In truth, any reliable source of combustibles could work.

What made the system more than the biotic analogue of placer mining is that the site would regrow, and the farmer would return. The second (and further) passes were easier, since trees would be smaller and more amenable to both slashing and firing. This "circulating" swidden, as it was aptly called in northern Europe, obeyed cycles set by regrowth and site conditions. Not every patch of woods or shrubland was suitable: swiddeners selectively targeted the most productive. As a result, landscapes were dappled with burned and unburned patches, with the burned swatches at various stages of recovery. It was a tamed version of what occurred in nature. Those patches also made a reservoir of biodiversity.

The effects of any single burn were short-lived. Farmers could plant in the ash and, in some cases, in the second year as well. But the indigenous flora (now identified as weeds) would return, and reburns could not continually renew the cultivars; the site would be abandoned or left to rough pasture. Farmers would move to a new site, perhaps one not previously used, or more commonly one that belonged to a cycle of return. This was a longer rhythm of fallowing that might extend not over a few years, but across several decades. (The Hindi term for the now-uncultivated land was *jangal* ["jungle"], which since the recharged flora was often dense, came to be applied to other environments like rainforest.)

Over a long period the character of the biome itself can change, and successive burns can result in regime change. Useful species endure, pesty species are weeded away. Such an outcome is obvious in places like central Europe, where fixed ownership and long-cultivated lands are typical; but it also characterizes shifting cultivation. Studies of the Ka'apor in eastern Amazonia have found that a large fraction of extant plants are utilized (perhaps 90 percent), and even those not directly exploited are "ecologically important"—surely, a result not only of clever learning, but of removing the undesirables. Equally, research demonstrates how much of the regional biodiversity is attributable to the low-grade swiddening that dapples the land with cut, burned, and abandoned patches of assorted ages and fluffs it with species not found in the untouched high forest.[9]

This style of fire-fallow farming was pervasive. It thrived in the mountains of Thailand, Finland's boreal forest and wet peatlands, the central hills of India, the tumbling terrain of the Philippines, the pine-steppes of Russia, North America's piedmont and coastal plains, Amazonia, Madagascar, the miombo woodlands of Africa, the denshired landscapes of England, the Ardennes woods of France—essentially everyplace it was possible and people had domesticated plants to cultivate. The terms for it are legion, and like Inuit words for "snow," they are often highly specific to landscapes and position within the cycle. Finns, for example, had different terms for shifting cultivation in forests, depending on whether the slashing was new or at a revisited site (*huutta* or *kaski*), whether it was a single-season episode (*rieskamaa*) or one planned over several years (*pyukälikkö*), whether it relied on local woods or on added peaty turf (*kyttlandsbruk,* from *kyteä,* meaning "smolder or glow"), and whether wetland burning occurred on eastern or western lands. Terms were as local and particular as the practices. Academics wanted a more generic term, something that sounded more abstract than *shifting cultivation* and less colloquial than *slash-and-burn,* and in 1950 adopted *swidden,* an old Norse word, long fallen out of use, to describe burning heather.[10]

Classic swidden required ample lands to roam among. A more intensive agriculture demanded closer attention and resulted from social systems that fixed ownership or held workers to a given plot.

Instead of the farm rotating through the landscape, the landscape (as it were) rotated through the farm. Plant succession came under the control of people as they planted one crop after another, each selected to take advantage of its position in the rotation, the duration of planting perhaps extended by weeding, until the site needed another fire. The land would be left to fallow, then burned, and the rotation would start again—another example of applied fire ecology. The rotation might be 2, 3, or up to a dozen years.

That worked for flora. Fauna required a different regimen and came with two particular concerns of its own. One pertained to fuels. Animals and fire competed for the same small-particle fuels—slow combustion in animals, fast combustion in fire. What livestock ate, the fire could not, which left pastoralists searching for adequate combustibles or the pastoral equivalent of fallow. In nature grazers could migrate to fresh forage and leave behind plants that (for whatever reason) were not eaten; even one season's passage made those plants far less palatable and unlikely to be revisited, and so they became available for burning. Among other ills, overgrazing stripped away that fallow-as-fuel, and the land lost the broad-spectrum ecological charge fire could provide. Together, plants, grazers, and fire made a three-body problem for fire ecology, one without an exact solution.

For societies that were primarily sedentary, there was a second concern—how to integrate flocks and crops, farmers and herders, and their fires. Europe provides some good examples. Two extremes required little or no burning. Reindeer herding relied on lichens, which are fire-sensitive, can take decades to recover after a burn, and argued against deliberate firing. (There was some useful forage stimulated by spring burns that supplemented that normal diet.) But since fires did occur in the boreal forest, the system needed large landscapes for the herds to roam around. The other outlier was the small farm that contained a tiny complement of livestock for special tasks, such as a cow for milking, an ox for plowing, a horse for riding. These were the faunal equivalent of a kitchen garden that could be intensively cultivated. The animals fed off fodder grown for them in the arable land, and they returned manure to the field as fertilizer.

But most arrangements required that the flocks move between arable fields (infield) and rough pasture (outfield). Infield and outfield might be parts of a single holding. Or the pasture might be some distance from the cultivated land or, if mountains were present, between valley and slopes. In Nordic Europe a saeter system took women and children to summer pasture in more distant woods or up mountains (mostly to produce milk for cheese that could be preserved for the winter). In Alpine Europe, hill stations were established to do the same. The land was burned to keep forage fresh and prevent woods from reclaiming it; and because the herders were families, if seasonally separated, social cohesion persisted. In Mediterranean Europe the ties were stretched. Large flocks moved to summer pasture in the mountains and winter pasture (often on stubble) in the valley, an arrangement called transhumance. Iberia devised a special arrangement that carried flocks long distances across the central mesa.

The outlying pastures were burned routinely; the routes and grazing lands of transhumance were annually fired, and the flames not always contained. The shepherds were themselves a separate breed and, like their fires, often poorly integrated with the socially ordered landscape. The Iberian versions were transported to the New World. Some herders, mostly Basques, took the tradition to the mountains, where pastoral fires became as much a seasonal feature as winter snow, while the long-ranging version morphed into the fabled cattle drives of the American West, for which arrangements might be made (most famously in the Flint Hills) for the land to be burned before the herds arrived, which meant both fresh forage and protection against wildfire.[11]

The agricultural ideal was a system that provided continual crops and animal goods within a coherent social system. Only a fraction of countryside could be cultivated or grazed, so mixed economies were the typical outcome. Farms might be located in valleys or floodplains, for example, with grazing in outfields or seasonal pastures, or hunting and gathering on uplands and remote sites; yet most uses looked to fire in some way—a soft ecological weld. And of course cooking—a second-order processing—was constant with fuels either grown or foraged.

. . .

Like aboriginal fire, agricultural fire had its limits. It required fuel that, for farms, had to be grown as fallow or brought in from surrounding outfields in the form of needles and branches, which meant that some land fell out of active cultivation for a year. Besides, only so much fuel could be coaxed or coerced out of the land. A first-contact bonanza in forest or moor could not be repeated except over long decades, if not centuries. Shortening the forest-fallow cycle too much left the site unable to recover sufficiently, leading to a slow wasting and impoverishment of soil. Slashing and burning, draining and burning, firing along the routes of seasonally moved livestock—all could expand the range of fire beyond its natural borders, but not infinitely or without cost. While the levees containing flame proved flexible, they could break if stretched too far. Pushing the land beyond its elastic limits led, as the saying went, to rich parents and poor children. If people wanted more fire power—and it seems they always do—they would have to find another source of combustibles. Europeans found one by expanding into new lands in the Americas and Australia, which, though inhabited, the newcomers regarded as fallow and used as such, but even that had limits. If they wanted more, they needed a larger and deeper reservoir of fallow.

In the meantime, as agriculture spread and interbred with aboriginal practices, humans' capacity to live on the land expanded. So did fire's. It burned in places and at times it could not have otherwise. Its pulses and patches were recoded to fit an agricultural almanac. Anthropogenic fire was not a precision instrument any more than maize or a milch cow was. It was not an appliance like a candle or stove, but it was as domesticated as the landscapes it burned. If humanity was touched by fire, so were the places that felt humanity's hand.

Pyrotechnics

Open flame was not the only pyrotechnology. Fire—the great interactor—enabled lots of other tools and practices that affected how fire would appear on landscapes. It was, indirectly, a force multiplier. In other words, it was not just the direct application of flames that projected fire's influence. The indirect consequences of pyrotechnologies were equally powerful. Humanity's fire practices had a long reach, eventually to such abiotic environs as Antarctica, the moon, and Mars.

Like fire itself, fire-based technologies had a strangely dual character, both physical and biological. Most devices were physical, the result of reductionist trial and error to isolate the heat and light that people wanted, with as little dependence as possible on open fire. The model was surely cooking. The hearth became a kiln or forge; the flame a firestick, candle, or blowtorch. The fuels could be refined into wax, spirits, or charcoal. Shaping the chamber for combustion could channel airflow—oxygen—and dispel smoke; eventually, bellows could assist. The commitment to metal tools or mining could, in turn, have major impacts on the woods used to fuel the forges and furnaces.

Fire could harden spears and help to shape flint. A torch could assist with night hunts and fishing—by freezing prey in the one, attracting it in the other. It could smelt ore into metal, then convert it into swords and plowshares. The axes it fashioned could slash woods, shovels could pare peat and drain wetlands, hoes could hack away weeds and lengthen the swidden cycle. Metal arrows and spears could assist hunting, which then influenced the fine fuels available for landscape burning, which was itself a kind of cooking.

Consider two practices that would seem removed from fire's Earthly ecology. Mining exploited fire to illuminate tunnels, to break rock, to smelt ore, and to fashion metal into tools. Fishing carried fire in offshore boats, used torches to draw fish to where they could be speared, and applied fire to cook the fish and smoke it to preserve it for the future. Both pyrotechs made parts of nature accessible to human artifice, and both, by sustaining people, further empowered humanity's realm.

Like a pyric double, fire showed all the attributes, for good or ill, of its alter ego. Like humans, it could fashion art, stimulate fields, and destroy cities. Not all pyrotechnologies were constructive and benign. Arson fire could destroy what craft fire created. *Fire and sword* was shorthand for war; *fire power* became a synonym for military might. Controlled fire was an enabler for whatever people wished, and their wishes were endless.

The Roman naturalist and encyclopedist Pliny the Elder wrote that "we cannot but marvel that fire is necessary for almost every operation." Fire made other tools and materials possible. It was the technical basis for alchemy and its scientific successor, chemistry. All of this enhanced the human presence. The architect Vitruvius separated the barbaric peoples from the civilized by their ability to convert fire to varied purposes. (The cooked and the raw as an index of culture continued into the twentieth century with French anthropologist Claude Lévi-Strauss.)[12]

Fire makes a peculiar technology. There are ample instances of it as a mechanical tool that isolates and amplifies some property or other. A candle bestows light; a torch, furnace, or dynamo, heat. Each identifies one property to promote a single effect; they can be turned on and off at will. But fire is also a biotechnology, a means to harness landscapes. Such fires derive their power from the character of their setting, and for complex ecosystems this means that effects are intricate and diffuse. Such burns can give us regenerating sequoias, blueberries, elk habitat, landscape mosaics—a diversity of sites and species. Physical pyrotech mimics hammers and levers; biological pyrotech looks more like sheepdogs and swidden plots, and in wild settings, it resembles a grizzly bear trained to dance. It is not a precision tool, but a broad-spectrum ecological process. Physical pyrotech simplifies fire's environment to ensure maximum control and a minimum range of consequences. Biological pyrotech derives its power from the complexity of its ecological context with outcomes that disseminate widely.

Historically, both had limits. Physical pyrotech was constrained by the availability of fuel, which came from living landscapes; biotech, by the character of those landscapes. Then physical pyrotech found a new world of fossil fuels that were for practical purposes unbounded.

Correlating Climates

Fire interacts with the atmosphere as much as with the biosphere. It affects air's chemistry and Earth's climate. The exchanges involve not just oxygen and carbon, but the planetary cycling of gases, especially those that promote a greenhouse effect. The rapid warming that ended the last glacial maximum removed ice and sustained a climate that could encourage landscape fire. The sapiens were uniquely positioned to move into the newly exposed landscapes.

In most parts of the world, anthropogenic fire was not something imposed on biomes that had evolved without it, but something added to a mix that already existed. This was most obvious for Africa, but as Eurasia rebounded from its icy burdens, sapient fire was there too, to coexist if not coevolve as biotas sifted and shuffled. Australia felt the firestick by 50,000 years ago. The Americas were the last of the vegetated continents to know humanity's fiery presence, but evidence suggests firesticks were used in Zacatecas, Mexico, by 26,500 years ago. And by 14,500 years ago, perhaps as early as 18,500 years ago, there were hearths at Tierra del Fuego. As the ice left North America, people moved in, or around, along the coasts.[13]

Aboriginal fire could not challenge climate, but it could work with it, nudge it in ways to promote more fire. Widespread burning in moister settings kept the landscape in grass or savanna, added to the regional fire load, regularized the fire regime fashioned by lightning, and kept carbon from being sequestered in forests that had returned exuberantly in past interglacials. Fire's impact was greatest where it was the primary mechanism for holding the woods back; it was most effective in the wetter grasslands such as the tallgrass prairie in North America, sourveld in Africa, cerrado and llano in South America, and the moister steppes of Asia. Fire's realm of influence could expand when the climate waxed warm and then be maintained when the climate cooled. Today, depending on how grassland is defined, some 20-40 percent of terrestrial biotas are in grasslands. Perhaps half are maintained in grass by burning or by burning and grazing.[14]

Agricultural fire further broadened humanity's effect on biomes, which is to say, on stored carbon. Forests were cleared, to be replaced by less carbon-hoarding flora such as grasses, shrubs, and young woods. Where the land was converted into wet rice paddies, methane—a greenhouse gas 14–20 times more potent than carbon dioxide—was pumped into the atmosphere. Similarly, pastoralism expanded the practices of hunting, and it too added methane, this time as a by-product of digestion in domesticated livestock. In 1961 an estimated 36 percent of the terrestrial biota was in agriculture. The number jumped to 39 percent by 1990 due to population increases and the conversion of Amazon forest to pasture and of Indonesian peatland to plantations. Then it dropped to 38 percent as a result of intensified farming, abandoned lands in North America and Europe, and the rapid industrialization of Asia, despite continuing pressures to expand land for farms and pastures. In the early phase of industrialization, countries increase production to satisfy increased population; in mature phases, industrialized countries reduce agriculture from marginal lands, expand cities, encourage reforestation, and create nature preserves.[15]

These are significant numbers. They imply that even under aboriginal management, a lot of carbon never got stored in forests and that under agricultural regimes, stored carbon was released and new sources of greenhouse gases were propagated. All this had to register climatically. The atmospheric outcomes of burning favored more burning. The expected cooling that ends an interglacial period stalled; the usual rhythms slowed and in some places reversed; about 6,000 years ago, the global climate stabilized.[16]

It didn't seem stable to those living on the narrow ledge of a subsistence economy, where seasonal and annual blips could mean the difference between a bumper crop and famine, but compared with the grand climatic swells that had typified previous epochs, the climate registered as anomalously steady. There were prolonged spells of warm and cold—the Medieval Warm Period (950–1300 CE) and the Little Ice Age (1550–1850 CE) are the most famous. The shock of the revanchist ice led to crop failures, unrest, ice-interrupted ocean traffic, and a general realignment of routine life with the imposed almanac of cold. But the apparent larger stasis is an unusual feature not characteristic of earlier epochs. In retrospect the latter half of the present interglacial appears as a "long summer."[17]

The conventional understanding has been to show how human societies adjusted to climate. Climate changes, people adapt. The warmer periods encouraged human settlement, so agriculture expanded. The cooler periods frustrated settlement, so cultivated landscapes receded. The Little Ice Age forced humanity into global retrenchment. But the relationship between climate and people is surely more interactive. If expansions involve land clearing, cropping, and swelling herds of livestock, greenhouse gases flow into the atmosphere. If human populations fall or settlements retreat, carbon is resequestered. At large enough scales, such movements can influence climate. Anthropogenic acts can stall or quicken the locked-in rhythms of the Milankovitch cycles. The Little Ice Age occurred after a catastrophic demographic collapse caused by plague in Eurasia and by European contact with the Americas that, over the sixteenth century, swept by a wave train of introduced diseases, caused a population collapse perhaps as great as 90 percent. Abandoned lands grew up to forests. The cooling ended when settlement pushed into new lands in the Americas, Siberia, and Australia and as a new source of combustion in fossil fuels became more widely available.[18]

These are historical coincidences that can serve as correlations but are, as yet, far from being recognized as causes. The record of carbon dioxide and methane in ice cores, however, is not hypothetical. It tracks shifts in greenhouse gases that align with climatic deviations from the patterns characteristic of previous interglacials. Either those shifts have a natural cause, not yet understood, or they follow from the ebbs and flows of humanity's presence. By this reckoning even the long summer was in part an outcome of human history: human finagling could not overwhelm the Milankovitch rhythms, but it could check and nudge them and grant an ever-flickering climate something like stability. The most tenable conclusion is that fire was not simply an epiphenomenon of climate, but in the hands of humanity, also an instigator of changes in it. There were limits since everything people did and burned resided within the domain of living landscapes and the gamut of outcomes that anthropogenic burning could effect. As with second-fire generally, people could push and pull the elastic borders, but not—not for long—exceed them.

Eventually the Little Ice Age faltered, then passed. Now another fire correlation has appeared. This one involves not living landscapes but lithic ones, biomass once alive but now fossilized into coal, gas, and oil. In effect, humanity discovered another new world, this one in deep time. It exhumed that fossil fallow, burned it in the present, and lofted its effluents into the future. This new fire did not burn openly in landscapes, but in special chambers that could isolate and amplify each of the components of combustion and deliver its energy to machines and locales removed from the place of burning.

The old ecological borders that had always confined fire dissolved. Like its fuels, this new fire—third-fire—was unbounded. And it was large enough not simply to tweak the natural mechanisms of climate, but to challenge them. Its gases filled the skies and, dissolved, dyed the oceans. Its flames, both the visible and the invisible, are filling the Earth.

4 *Fire Creature*

Lithic Landscapes

James Boswell, diarist extraordinaire and biographer of Samuel Johnson, is not recognized as a figure of note in fire history, and he wasn't. But in 1776 his life's journey intersected fire's journey in curious ways that rendered him a chronicler of not just Britain's great Augustan but how his times altered the trajectory of fire history.

In late March he and Johnson traveled to Birmingham. Later, Boswell trekked 2 miles outside the town to Soho Manufactory, there to visit Matthew Boulton, the business partner of James Watt and the fabricator of the Boulton & Watt steam engine. Boswell lamented that Johnson had not joined him, because "the vastness and the contrivance of some of the machinery would have 'matched his mighty mind.'" A member of the Royal Society and the Lunar Society, one of the founders of the Industrial Revolution, Boulton stated his business directly to Boswell: "I sell here, sir, what all the world desires to have—POWER."[1]

On April 2 Boswell, again without Johnson, attended a dinner with Sir John Pringle, then president of the Royal Society, and Captain James Cook, who had returned from his second voyage around the world. Boswell related how "while I was with the captain I caught the enthusiasm of curiosity and adventure, and felt a strong inclination to go with him on his next voyage." Johnson expressed doubt that one could learn much from such travels. He had translated Father Lobo's *Voyage to Abyssinia* and written a didactic novel, *The History of Rasselas, Prince of Abyssinia,* which advanced the same argument. But the era agreed with Boswell, and its winds filled the sails of what became a second great age of European exploration and colonization. Three months later Britain's American colonies declared independence. Europe's settler societies would become imperialists as well.[2]

The steam engine could run on wood, but it soon exhausted reserves and turned to coal. In fact, among the earliest uses of Watt's engine (as for Newcomen's before him) was to drain water from coal mines. Like a self-reinforcing dynamo, steam power was used to exhume more fuel to create more steam power. At the same time that the steam engine was deconstructing fire, the Enlightenment—in which the Royal Society was an active agent—was removing it as a phenomenon in its own right, dividing its parts among physics, chemistry, and mechanical engineering. As fire disappeared into engines, so it vanished from intellectual life. It was no longer seen as a universal principle, or a transcendent means of testing and explaining, and instead diffused into the concept of energy. It became a subject without a disciplinary home, an intellectual foster child handed around relatives, some more accepting than others.

Meanwhile, James Cook commenced his naval career by transporting coal from Newcastle, and he later retrofitted a coaler into the HMS *Endeavour,* in which he first circumnavigated the globe. Among the consequences of that fabled voyage were the discovery of eastern Australia and the circumnavigation of New Zealand, both destined to become sites for British colonization. While wooden ships and sail were enough to export people and rule over lands and waves, the new era could rely on steam and steel to cross both oceans and continents. This second age of discovery became a vector for exporting the Industrial Revolution around the Earth.

All in all, it was, as Boulton bluntly put it, about power. This time humanity's fire power was unshackled from its biotic context; then like all fires, it began remaking the world to promote more of itself. The ancient world had considered fire one of four elements. In Greek myth Prometheus had ripped it from its divine setting to bestow it upon humanity. For his defiance he was shackled to a peak in the Caucasus Mountains; and so, too, his gifted fire remained fettered to larger ecological processes. In 1818 Mary Shelley wrote *Frankenstein; or, The Modern Prometheus* that told of a scientist breaking the established biological order, and 2 years later her husband, Percy Shelley, wrote *Prometheus Unbound,* a lyric play that celebrated the liberation of the titan. So, too, the new fires were liberated from their ancient fetters, passing over the planet like an invisible flame, transmuting whatever they touched. The century had begun with the first steamboat and steam locomotive; it ended with unbound modern Prometheans refashioning the Earth through industrial combustion.

Pyric Transition: A New Order of Combustion

This transformation—call it a pyric transition—began as fossil biomass, notably coal and then oil and gas, selectively replaced living biomass, mostly wood and peat, in various fire appliances like hearths, forges, and furnaces. Over millennia people had devised ways to refine fuels and find suitable apparatus in which to burn them. A taper burned wood; a candle, wax; a lamp, kerosene from whale blubber. A hearth had a flue to control airflow; a forge, bellows to force air. It was relatively easy to replace wood with coal in both. A village of hearths or an abundance of forges or smelters for metallurgy could quickly strip a landscape of its woods. Relying on wood meant people had to spend more and more time foraging for fuels or else move, and they did both.

Fossil fuels changed this dynamic—in the most radical reformation of anthropogenic fire since the acquisition of the firestick. The shift required a suitable combustion chamber; this first emerged with the steam engine over the course of the eighteenth century. The engine could then be adapted and distributed widely. With human ingenuity and promethean ambition, this new fire could, like the old, propagate. Its advent marks a phase change in planetary fire history.

Coal has a higher caloric content than wood; anthracite has more than twice the gross caloric value of dry wood. And natural gas has more than a 50 percent greater value than coal. Fossil fuels are abundant—Earth has known terrestrial life for a long time and has laced the stratigraphic record with its biomass. Mined ore was not uniformly distributed, and because it was bulky, those places that were near deposits were the first to use it at scale. The new fire engines made possible more engines, which both helped make them accessible and demanded more fuel. Additional sources of fossil biomass were found—vast reservoirs of untapped fallow like discovered new worlds from the geologic past. More machines, more fuel—the process became exponential, and humanity's fire power with it.

If the amount of fuel seemed unbounded, so was its burning. Living landscapes had ecological boundaries and internal checks and balances. Combustion of biotic fuels tracked the availability of combustibles, and these were set by weather, seasons,

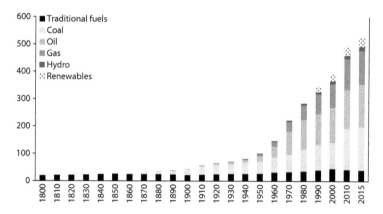

FIGURE 2. Global primary energy consumption from 1800 to 2015 as measured in exajoules (EJ; 1 EJ = 10^{18} joules). All energy sources listed except for hydro and renewables rely on combustion (93%). Data source: Vaclav Smil, *Energy Transitions*, 2nd ed. (Santa Barbara, CA: Praeger, 2017).

decomposers, and the phenological cycle of greening and curing. In carbon as in ecology, fires tended to be conservative: what flames released, the renewed biota recaptured. People could tweak those conditions—could use wind to guide fire and disperse smoke, could slash and drain places that would otherwise resist burning—but they could only coax and cajole so much out of a landscape before its elastic limits were exceeded and its capacity to grow fuels degraded.

Lithic landscapes knew no such boundaries. They were mined, not grown. They could burn day and night, winter and summer, through drought and deluge; they could burn in stony desert and rank rainforest. Their fires did not recycle carbon but transferred it across deep time. Their fires were limited only by the ability of humans to exhume and transport fuel, and as the fire engines ramified and spread, so did the capacity to gather more fuels to burn.

The old quest for fire had been a quest to find ever-more combustibles, a search for fresh sources. The new challenge was to find places to store all the effluent (and deal with the knock-on effects) that the burning produced, a search for sinks. Within a couple of centuries, the new combustion filled the atmosphere, percolated into the oceans, and grew back on lands. The atmosphere warmed and unhinged the old climate. The oceans acidified and rose from melting ice sheets. The lands thickened with new woody plants, adding fuel to living landscapes. The ancient constraints within which fire had to burn broke down. Power plants' capacity to produce exceeded their capacity to absorb the inevitable by-products; they were factory farms for combustion.

The new regimen of combustion also changed the relationship between fire and people. Their fire power shifted from direct application to an indirect impact, mediated by machines or through products made possible by fossil biomass. Fire with its flame, glow, heat, light, and crackle was reduced to the most elemental chemical and physical expressions, each isolated and engineered, so that what had been "fire" became "combustion," and combustion had become only its constituent parts. Lithic landscapes interwove with living ones. In Howard Odum's famous formulation, "industrial man no longer eats potatoes made from solar energy; now he eats potatoes partly made of oil." Instead of using fire to grow feed for draft oxen, people could plow with tractors fed diesel. Instead of flame to fumigate and fertilize, they could appeal to herbicides, insecticides, and synthetic fertilizers derived from reserves of fossil biomass and processed by fossil-fuel-burning machines. The material world filled with plastics and energy pathways, both drawing on a reservoir of fossil biomass. And of course the same applied to built environments. Instead of lighting a room with candles and hearth, people could illuminate with light bulbs powered by electricity generated at coal-fired power plants far away.[3]

The new order distanced people from fire. They no longer held it in their hands or saw it in working flames all around them. Its presence had been sublimated; the sense of fire as something people, uniquely, manipulated seeped away. The very presence of fire vanished, save as disaster or something sequestered, like wolves and grizzly bears, into remote sites. Fire disappeared from intellectual inquiry; it was reduced in the best scientific and technological fashion and bundled off to the academic equivalents of the new appliances. Much as fire was parceled out to a menagerie of machines, so its study dispersed among disciplines. Combustion went to oxidation chemistry; heat, to mechanical engineering; light, to electromagnetism. In a bizarre way, fire had helped deconstruct itself not only as a natural phenomenon, but also phenomenologically, as a concept.

Paradoxically, humanity's escalation in fire power came at the expense of real fire. The new order actively sought to eliminate open burning by technological substitution and suppression. The replacement technologies ran off fossil fuels and electricity. Suppression meant stopping people from using fire and then applying mechanical force powered by fossil fuel combustion to extinguish fires that did occur. Remove pumps, chainsaws, engines, airplanes, helicopters, trucks to carry crews, bulldozers to plow fire lines, and the roads made by graders and dozers, and it would be impossible to close-contain fires in rural and wild lands. This strategy makes sense in cities, but not in the countryside. In the past, fire officers had relied on preemptive burns and on emergency burnouts that would leave extensive fire on landscapes. As these shrank with the acceleration of fire-suppressing machinery, the loss of fire left a combustion deficit, an ecological mess, and a stockpiling of living-landscape fuels. Before long, those fuels would exceed the capacity of combustion engines to contain the fires that consumed them. Even emergency burnouts and prescribed fires relied on drip torches fueled by diesel and gas, on incendiaries dropped from aircraft, on fire lines cut by dozers and graders and protected by engines, and on crews transported to sites by gas-powered vehicles.

As the first-fire of nature had yielded to the second-fire of humanity, so second-fire ceded to a third-fire, and second-nature segued into a third. Second-fire was nature's fire directed by people; it powered the Anthropocene, but if people left, living landscapes would still burn. Third-fire was solely a human construction and would expire if people quit tending it. But it put the fiery remake of Earth on afterburners. The energy source behind the Anthropocene so dominated everything else that the epoch could more rightly be called a Pyrocene.

Pyric Transition: The Concept

The turning inside out of humanity's relationship to fire deserves a name. *Industrial combustion* seems too bland, though no more so than *aboriginal fire* or *agricultural fire*. But the much-studied process of industrializing suggests an option for the moment of phase change, the kindling of this new fire as it were. Among the changes that the general transformations set in motion by industrializing is a "demographic transition" within the human populations experiencing it. Perhaps something similar occurs with the "population" of fires. In its original incarnation this was the expectation.

Human demographics combine two separate ratios or trends—the birth rate and the death rate. Initially upon industrializing, deaths decline but births continue. Then births decline, even below replacement value. For decades the overall population remains high because the old generations linger, but eventually the smaller new generations pull the cumulative numbers down. As countries begin industrializing, they suffer population explosions; in mature countries, populations decline. Something similar seems to happen with landscape fire.

As we should expect in a new form of fire colonization, the shock wave expresses itself in many ways that depend on the particulars of the firescapes it encounters. In humid forest lands it leads to an explosion of burning, as axe and fire strike lands opened by, and connected to global markets by, fossil fuel transportation, with abusive and often explosive burning the outcome. Fire's presence blows up as old practices continue and new fuels and ignitions become available. Then, with decades of substitution and suppression, it diminishes, dropping below replacement value, or the ecological requirements for burning. In grasslands and arid lands, the process leads to the reverse. Burning implodes because those same arrangements encourage replanting to commercial crops, or overgrazing, or both, as former combustibles are replaced or fed to cattle and sheep. The effect is nearly instantaneous. Fire disappears from the land, perhaps returning only as woody vegetation or invasive exotics replace native grasses. Again, the needed population falls below ecological replacement value. To this extent the demographic transition serves as a heuristic concept.

The analogy can turn spongy, perhaps more metaphor than model. What exactly needs measuring? Is the population of fires the right metric? Or the area burned? Or is it the carbon throughput that combustion processes? The analogy falters, too, in that it is the death rate rather than the birth rate that declines. Traditional burning—

second-fire in all its iterations—withers away; fire famines spread among many biotas. Meanwhile, the new burning—third-fire—continues exponentially. The Earth has more combustion than it can absorb. There are no internal checks and self-corrections comparable to the local, individual decisions to limit family size that shape human population.

The transition manifests itself locally on the land as people cease traditional burning, look for alternatives, and suppress fires of all kinds. The process passes across the Earth like the terminator. Its affects are first patchy, displayed by society, by nation, and by region. But with time it also globalizes through its impact on trade routes, on ideas and institutions, and on the atmosphere. Commerce, colonization, and applied science can bring the effects of the transition to places that have not, on their own lands, experienced it. Climate change can affect places that have not themselves undergone a pyric transition on the ground. Overall, fire stimulates more fire. The process deserves targeted research; China and India are rapidly industrializing and offer fresh data, while Indonesia (in Borneo) and Brazil (in Amazonia) show the transformation in lightly inhabited lands.

An explicit model may not matter so much as recognition that living and lithic landscapes are joined by fire. Yet like the demographic transition, the pyric transition manages to illuminate a global theme and show how a common cause can result in different outcomes for two processes. That is not a bad approximation of what has happened with Earthly fire. At first, at particular places, second- and third-fires compete. Only where fossil fuel combustion actually affects how people live on the land does it challenge second-fire. Countries may hold coal or oil in abundance but export them or divert the wealth into elites (the resource curse). With time, however, the globalizing effect of greenhouse gases reshapes the climate so that the transition affects every place no matter how removed from the quotidian realm of second-fire or, for that matter, fire at all. Coal plants in China, India, Germany, and Pennsylvania can melt the pack ice of the Arctic Ocean, the glaciers of Iceland, and the ice sheets of Greenland and Antarctica. They can influence fire regimes in Australia, Argentina, and Siberia.

Taking fuels out of the geologic past, burning them in the present, and releasing their by-products to a geologic future—this is fire's new narrative arc and one of the grand markers in Earthly history. How those two realms of fire—living landscapes and lithic landscapes—interact in the present is the pyric transition. To date, it has been little studied in any systematic way, and little understood as an integral theme of fire's ecology. If considered, fire is seen as an epiphenomenon, not an essence. It is typically used to promote alarm over shocks like climate change, rather than identified as the mover behind those changes and the narrative driver behind their understanding.

Pyric Transition: The Practice

Given time, the dark passover that is the pyric transition affects every environment that humans inhabit or just touch. By technological substitution it replaces fire with surrogates and transfigures flame into electricity, changing the character of combustion and humanity's relationship to it. By empowering humanity on a gargantuan scale, it has upset biogeologic cycles of carbon, nitrogen, and sulfur and allowed people to ship bulk goods like sand and garbage across oceans. By influencing how people organize landscapes, it alters the fundamental conditions, especially fuels, of fire's environment. By infiltrating, then flooding the atmosphere, it sways climate, and through climate, the entire surface of the Earth. From the desktop to the globe, humanity's fire power is remaking the planet. Third-fire is creating a third-nature.

The transition is removing working flames from every human habitat. It takes fire out of houses, out of factories, out of cities—the built environment. It erases fire out of aboriginal and agricultural landscapes. It strips flame even from sites remote from permanent settlement, from recreational parklands, and from nature pre-serves. It resembles traditional patch burning in reverse—fires expire in local places that, over time, mass into flame-free dark-ness. As humanity's new fire power reshapes transportation and climate, the patches merge into larger expanses, some approaching continental scales. Even in satellite imagery, fire's removal has become as conspicuous as its presence. The transition explains the grand paradox that even as wildfires rage across screens and head-lines, the amount of fire in living landscapes is declining.[4]

Or rather, fire's attempted removal. In built environments suitable replacements are possible and usually desirable; flame becomes ceremonial. In natural settings prone to fire, however, even its attempted suppression can disrupt and worsen the fire scene. Humans had devised a new tool and then used it on every-thing without thinking about long-term consequences. The fact that fire—other than as a tool—had more or less vanished from advanced Enlightenment societies as a serious object of inquiry made it easy to ignore the probable outcomes of removing it from landscapes as one might remove it from a kitchen. A new ecology of industrial combustion emerged around the dynamics of a menagerie of machines.

Third-Nature's Built Landscapes

The transition has worked most completely in built environments. Once, houses and towns were constructed of the same materials as the surrounding landscape; even where brick, adobe, or stone was used, the roofing was made of lighter materials, which came from the combustible countryside, or was supported by wood. Such towns burned as often as the surrounding environs.

Now urban cores are made of incombustible materials, which have in a sense already passed through flame to become steel, glass, concrete, or brick. Where interiors were once filled with working flames—candles and chandeliers for lighting, hearths for heating, stoves for cooking—these flames are gone, or they have yielded to more-controlled fossil fuels like natural gas or propane. Light bulbs have replaced candles; electric ranges or propane stoves have replaced cooking fires; electric, gas, and fuel oil furnaces or heat pumps have replaced the glowing fireplace. Interior furnishings are tested for flammability. Even fire protection depends on electricity and gas to run alarms, trigger sprinklers, power pumps and engines, and supply the infrastructure. The days have passed when cities had to suffer routine conflagrations or to counter flame with bucket brigades and hand-powered pumps, or to tear down structures to create a firebreak.

Yet the transition to a fossil fuel civilization has ironically assisted the return of fire to built environments by encouraging a new kind of urban landscape, one that put the built and the wild into sufficient proximity that fire could span the gap between them. In the United States between 1992 and 2015, a million houses lay within a wildfire perimeter, and an estimated 97 percent of residences outside an urban core were potentially at risk from wildfire.

This was a result of how Americans lived on their land—how transportation allowed service economies to recolonize a formerly rural scene or how urban structures could abut wild countryside or, more broadly, how living landscapes could interact with lithic ones. Internal combustion encouraged dispersed settlement, fallow lands shrank, power lines cast sparks. Each action had its particular reaction, each cause its effect, but the emerging crisis was the result of how the various constructions of a fossil fuel civilization were combining in unexpected and dangerous ways.[5]

There is much to commend in the transition. Indoor cooking fires and smoke were a chronic health hazard, and heating fires could put cities under a wretched pall. Cities no longer burn in conflagrations: it takes an earthquake, war, or riot to shatter the cityscape into burnable debris and break down the capacity to respond. The threat of fire persists—like a strange attractor, unseen but deforming its setting; but fires are becoming rarer. Most modern fire departments have evolved into all-hazard response services for which fire is a minor contributor. The third-nature of third-fire even has faux fires or films of fire that blaze on monitors. (Instead of burning a Yule log, people can play videos like *Fireplace: The Movie*.) A home entertainment center replaces the hearth.

Yet some features persist. Traffic lanes are the new lines of fire. Power plants, malls, factories, apartments, and any places where people gather and demand energy are new fields of fire. Though fire is absent, its omission is by design. A modern building will be made of materials tested for flammability, will have smoke detectors and automatic sprinklers, will have capacity determined by the need to evacuate in case of fire, will have fire exits marked by signs complete with their own power source. Even in its absence fire shapes industrial cities.

Third-Nature's Rural Landscapes

What surprises is that a similar scenario applies to the second-nature world of second-fire. Here fire is not simply a mechanical tool, an appliance for exploiting one or another property of fire, but a process that results from interacting with an open-air, lightly controlled landscape. Its properties derive from the character of that setting: it is semitamed, like a fallowed field, the postharvest stubble of sugar cane or wheat, the slashed woods of a swiddened milpa. It less resembles a windup clock than an elephant or horse captured from the wild and trained to harness. Control has its limits, dependent on the character of the prepared site, only parts of which can be manipulated with anything like the kind of precision of a fire-powered engine.

What farmers and herders want is the ecological jolt provided by fire. From a flame's passage they get a restructured microclimate, a soil both stimulated and purged, nutrients that were locked up in dead wood or debris that are liberated into ground, water, and air. A good fire temporarily purges the old biota and readies the site to promote new cultivars or refreshes forage for a new cycle of grazing. Agriculture requires a variety of tasks: fire performs many of them in one integrative wave.

But the same process of reductionism that can turn a campfire into an electric range can contrive alternatives to burning fields. Fire transmutes stubble and slash into accessible nutrients and typically stimulates nitrogen-fixing organisms. The alternative? Manure and artificial fertilizers. Fire can burn away the unwanted vegetation at least long enough to plant a year or two of crops. The alternative? Chemical herbicides and tractors. Fire sweeps away and smokes away, for a time, problem species. The alternative? Chemical pesticides. Fire does all this in a single process of flaming and smoldering. The alternative? There is none. In intensive cultivation, there is no desire to achieve all these outcomes, only those few that maximize production.

Fire without flame, smoke without fire, fertilization without ash, fumigation without smoke—industrial agriculture selects for what is needed and discards what isn't. Each aspect of agricultural fire can be identified, isolated, and enhanced; that, in fact, is what the scientific method is designed to do. An integrative process like fire does a score of things, none of them maximized. In principle it should be possible to enhance each separate effect. Much as a light bulb yields more light without more heat, or an electric range yields heat without smoke, so can technological prowess be applied to agriculture and ultimately strip fire out of fields or, more properly, replace flame with more specifically targeted processes. Artificial fertilizers, pesticides, herbicides, karrikinolides to prompt flowering—each can replace a feature of fire. Their delivery systems are machines such as pumps, tractors, and aircraft powered by fossil fuels. What fire had done in one sweep of flame and smolder, a suite of techniques now does, each specific to a single task.

The goal is to maximize the effects of separate features and ultimately the economic productivity of the land. Unlike a factory, however, engineers can't control all the elements, so it is impossible to reassemble each enhanced part into a comparable system. There is no mechanical or chemical surrogate for fire in all its integrative complexity. Fire-fallow agriculture had tweaked first-nature into a second-nature but had retained most of its parts and their interactions and had used fire—also part of first-nature—to assist the transition. Third-fire agriculture distributes the chemicals without the species that deliver them in first- and second-nature, and it rearranges the scene without the catalytic jolt of fire.

The farm more closely resembles a manufacturing facility, and like industrial factories it dispenses with fire in favor of machines that can each perform a single task without the complex interactions favored by fire.

With no apparent need for fire, there is no obvious need for fallow, which European agronomists had long loathed. More of the land could be cultivated, mostly for single crops. The patchiness of swidden or rotational farming, with its mosaic of slashed, plowed, planted, and variously fallowed sites, disappeared. And with it vanished a magnificent matrix for biodiversity along with the many species useful for foraging, medicines, and supplemental hunting and trapping. Second-nature had reorganized first-nature to better suit human ambitions; third-nature simplified the scene by eliminating sites not geared for maximizing production. Habitats vanished, and where uncultivated sites survived, they did so without the fires that powered the old arrangement.

But like fires that deconstructed into engines, fallow found a new identity. Those vast reservoirs of fossil biomass served, for third-nature, as fossil fallow. Not only did they furnish fuel, they were a source of chemicals that, when smelted, became fertilizers, biocides, and the raw mass from which to create alternatives to such natural materials as wood, stone, and metal in the form of plastics. Humans expanded into a new world, not one of land but one of resources made abundant by bringing fossil fallow into production.

Third-Fire's Wildlands

Humanity's new fire power first boosted the conversion of forests and grasslands to farms and pastures, then by intensifying agriculture, effected the loss of fallow and a withdrawal from marginal lands. By laying down transportation systems capable of connecting resources, however remote, to consumers, it also encouraged market hunting—think bison for hides, and bird plumage for ornamental hats. Accordingly, state-sponsored conservation took on overhunting and loss of habitat along with forest destruction as part of its core mission. If nature did not impose limits, at least at a meaningful timescale, humans would have to.

Forest reservations were one response; parks and nature preserves were another. Together they created an alternative landscape: these were to third-nature what fallowed lands were to second-nature. They were not just sacred groves, or hunting preserves for kings and aristocrats, but significant chunks of countryside under state (public) administration and dedicated to retaining first-nature—to prevent the full consumption of natural landscapes by the terminal forces of industrial combustion. The growth of protected lands matches the curve of fossil fuel combustion, and while correlation does not mean causation, it is hard to believe that the emergence of nature reserves of various sorts was not a by-product of the pyric transition. The economics and esthetics of the preserves are those of industrial economies. They were a reaction to the shock wave of the pyric transition.

Most reserves burned, so the question arose of how to respond to their fires. It was not inevitable that industrializing societies should apply the same expectations for fire in wildlands that they applied to cities. The fact that they did—and established a mistaken template for the future—is the outcome of various factors that converged with the transition. None were truly independent, since all were themselves magnified and brought together with the assistance of third-fire.

Begin with Europe's renewed expansion, this time by northern nations. Some countries, like India, Algeria, and Ghana, Europe ruled, if by indirect means. A few received massive immigration and became settler societies. All felt the quickened pace of a globalizing economy, or at least the looting of forests, wildlife, soils, and ores by a footloose capitalism. Steam cracked open interior landscapes and made possible shipping for a world market. People didn't need railroads to fell forests or steamships to haul wood across oceans (or in prairies and savannas, to overgraze to nubbin), but industrial transport connected suppliers to consumers at a pace and scale unthinkable previously. Mass overgrazing in the American West, clear-cutting through the northern forest, sod-busting across the Great Plains—all could not have occurred at anything like their pace, intensity, and extent without the catalytic presence of railroads.

The wreckage of forests especially led to outbreaks of conflagrations. Local governance proved inadequate to halt the rampage. The state, notably an imperial or national government, would have to intervene—would have to stand between the woods, waters, and soils and the loggers, miners, graziers, and others who, outfitted with unmoored capital and the transformative power of steam, swept over public or communal lands and left huge, often combustible messes behind. In the name of state-sponsored conservation, it did. This was the axiomatic rationale behind extensive forest reservations and forestry bureaus to oversee them. Their primary task was to regulate logging and suppress burning—to spare the national estate from the havoc of untrammeled "fire and axe." (The rule of thumb was that wildfire destroyed ten times as much forest as logging.) The project was intended to spare the present generation from disasters, and future generations from "timber famines" and wrecked watersheds.

All this was what modern, rational states were supposed to do. Similar institutions and ideas spread across the French, British, Dutch, and Russian empires and acquired second lives in settler societies like those in the United States, Canada, and Australia. Germans excelled at forestry; the French bonded it to the ambitions of the state; and the British established the model of forestry in the service of empire. There were economic interests, of course, but forestry was also a justification for taking over lands from peasants, Indigenous peoples, and less Enlightened states. Interestingly, a major consideration was climate. Early colonizing of islands had impressed on Europe's savants that deforestation by axe or fire would lead to destabilized weather prone to droughts and flooding.[6]

The reserves were turned over to foresters to administer. It was a case in which the name mattered: of course foresters should oversee forests. But the reserves had purposes beyond timber, and forestry was probably the worst-equipped guild to intellectually cope with fire. It not only hated and feared fire; it knew nothing about it. It considered fire within the context of temperate Europe, one of Earth's anomalous lands that had no natural basis for routine fire. What fire existed burned at the hands of people, which suggested that fire was a social problem. The founders of state-sponsored forestry, like Bernhard Fernow, even insisted that fire was not a part of formal forestry, that fire control was a precondition to forest management. Fire control, much less anything that might approximate fire management, was not even in forestry's professional curriculum.

Yet overwhelmingly fire protection became its defining task and an index of its success, and for most of the twentieth century, foresters sought to eliminate fire as fully as possible. They discouraged traditional burning, they suppressed fires from any cause, and they pursued smokes to the most distant hinterland. As late as 1953 an American text on fire control observed that "the effect of this [traditional training] has been that many young foresters have found themselves on a job of which four-fifths was protection of the forest from fire, but with their training in inverse ratio." If fire hijacked the traditional concerns of European forestry, it also made forest bureaucracies powerful. It gave them an enemy to fight, a marker by which to measure success, and a visible presence before the public and their political masters.[7]

Because the great reserves were administered by foresters, fire science emphasized control, and because foresters became, if by default, the oracles and engineers of landscape fire, their doctrines propagated beyond the reserves themselves. Many wildlife reserves, mostly established to preserve big game, came to a different perspective, that fire was needed even if it seemed objectionable; yet they fell under the common rule of forestry's fire-suppression project. So did national parks, from the United States to Brazil to Ethiopia, all of which sought to rein in both human burning (which was unnatural) and lightning fires (which were intrinsically damaging). Thanks to forestry's impressive political power and institutional authority, its rule even extended to the rural countryside generally. Forestry agencies became the face and force behind national fire protection outside cities. There was no academic or scientific counterweight to challenge forestry's interpretation of fire. The shock wave of flame triggered by the tremors of the pyric transition continued well after the era of conflagrations had passed.

Early-onset fire control proved relatively simple (though it didn't seem so to many practitioners). In the early years fires were easy to control on fire-frequented landscapes because ceaseless burning had kept combustibles low or, in places prone to crown

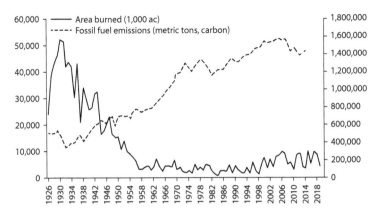

FIGURE 3. Two fires, competing: US burned area (in 1,000 acres, left axis) and fossil fuel emissions (in metric tons carbon, right axis). Fossil fuel emissions source: Carbon Dioxide Information Analysis Center, US Department of Energy. Burned area source: National Interagency Fire Center. Data for 1926–70 come from *Historical Statistics for the United States*. States are included as they joined the Clarke-McNary Program; Alaska, since 1959. Common reporting forms apply only since 1983. The upshot is that more areas are included in the counting over time, but even so, the area burned declines. Pre-1926 figures for area burned are almost certainly higher than the 1926–30 tallies. The data pertain to wildfires and do not include agricultural burning, which has continued but at declining rates.

fires, had kept the canopy patchy. Fire engines and the inherited landscape meant fire control could be successful for most fires. With time, as fuels spread and thickened, fire control demanded yet more power to meet the more forceful flames. Eventually, only extreme fires were left—precisely the ones that did the most damage, cost the most to contain, and were uncontrollable until winds or fuels died out.

A revealing story unfolded in Northern California in the early twentieth century. Two powerhouse US Forest Service personalities,

S. B. Show and E. I. Kotok, described the situation that had evolved. When the national forests were established, everyone had been engaged in forest burning, which was an "established practice." The idea that "fires could be excluded entirely from millions of acres was generally regarded as preposterous"; critics prophesized all kinds of ills were it to be attempted, not least "the uncontrollable crown fire." As fire protection strengthened, the amount of flammable material, exactly as forecast, "increased greatly." That prompted locals to take to fire—some as arsonists, some as protesters attempting to reinstate traditional burning, but both regarded equally as threats "not only because of their direct action, but even more so because of their open preaching of fire." In the end, forestry triumphed; foresters got the thick regeneration and scrub that a later generation would identify as stoking uncontrollable wildfires. There was, at the time, not the slightest sense of irony or doubt among the authorities. Between 2017 and 2020, precisely this region was hit by serial conflagrations, one that burned into Santa Rosa, one that spun into Redding as a fire tornado, and one that simply burned down the town of Paradise with a loss of eighty-five lives.[8]

Parks, wildlife refuges, and pristine nature preserves felt the impact not only in the form of worsening bad fires, but from the lost ecological benefits of good fires. By the latter twentieth century, forestry fell into disgrace; fire's restoration, not its exclusion, became the goal of management. Still, among developing countries eager to advance, fire suppression remained a powerful emblem of modernity—a fight not only against flames, but against traditional practices that had retarded development and left countries mired in superstition and backwardness. Originally, elites had opposed the folk and argued for fire's abolition; now, in mature developed countries, the elites wanted to reinstate fire while the public, long subjected to antifire messaging and now mostly urban, favored suppression. The ironies flow back into themselves like a Klein bottle.[9]

Here is the so-called fire paradox. What worked in cities failed in wildlands. Through a series of historical accidents, or convergences, the pyric transition had disrupted fire scenes such that there were too many bad fires and too few good ones. In Australia an effort to base fire management on controlled burning emerged in the 1950s; in America, during the 1960s and 1970s. As the full ramifications of fire exclusion became apparent, forestry was forced to surrender its monopoly over fire science, and then over fire management. The decolonization process that boiled up after World War II not only liberated countries but decommissioned the bureaucratic instruments of imperial rule, forestry bureaus prominently among them. The legacy of the era, however, was baked into those disrupted biotas; populations were mostly urban, removed from the habits of quotidian fire; and traditional fire cultures were isolated and feeble. Efforts to restore fire would struggle.

Yet not the least of the paradoxes that spill out of the Pyrocene is that much more fire is to come. The only choice is whether it will be wild, feral, or prescribed.

Big Burns

The pyric transition begins with competition between fires but segues into collusion among them. Satellite images of Earth at night illustrate the competition nicely. There are two distinct realms of light. One emanates from flames in living landscapes; the other, from illumination energized by combusting lithic landscapes. Because the gaseous effluent gets mixed in the atmosphere, all those separate sites and sources can affect settings everywhere, even those that have not made the transition. A warmed globe then acts on fires in living landscapes as a performance enhancer, most often in ways that quicken, broaden, and intensify their fires.

Fires worsen in wildlands. They burst into exurbs, occasionally carrying flame into urban cores, from Santa Rosa, California, to Gatlinburg, Tennessee. They interact with land clearing, logging, invasive species, human-based ignitions, and an unsettled climate to convert landscapes, usually into more fire-prone avatars. The cumulative burning drives out the last vestiges of the Pleistocene into refugia, and then to extinction—its ices, its megafauna, its fire-intolerant flora.

The transition is lumpy in space and time. It races through some places in decades, while in others the effects are felt indirectly through climate, rising seas, plagues, insect infestations, invasive exotics, and native extinctions. Nor is the transition linear and methodical, substituting one variety of combustion with another, like silica replacing lignin to make petrified wood. The pyric transition is not a substance but a process. For humans it may take several generations; in other than human time, it can happen abruptly, like a hydraulic jump in the river of time. In one sense humanity's fire power has been strengthening since the last great ice sheets began to recede. Third-fire gave afterburners to the propulsion behind the Pyrocene.

There is much to commend in how the pyric transition has remade humanity's built habitats. Kitchens and towns are not filled with smoke, fire no longer routinely threatens combustible houses and cities, and there is no need to forage constantly for fuelwood. Unlike a child or second-fire, third-fire does not need ceaseless tending by much of society. The transition, once past its shock wave, abusive phase, can be a sweet spot for human life. The drudgery of fire tending disappears. Much as kitchens in warm climates were often separated from the main house, so now the source of power exists away from its place of use; and instead of cloying smoke, invisible clouds of emissions flood the atmosphere. The dangers of open fire, especially in built environments, diminish.

But the pyric transition comes at a cost. The exchange works at a local level: one kind of fire or the other exists at any one place. Similarly, avoiding the messiness of flame in agriculture leads to chemicals that can pollute. (Nearly all fire retardants can be carcinogenic, reminding us that burning is basic to the living world, not an alien imposition.) Banishing fire from wildlands and countrysides that have adapted to it can destabilize landscapes and encourage bad burns. Like the discovery of antibiotics, which have been overprescribed and misapplied and rendered useless, the overuse of fossil fuels has created conditions that have made it possible for wildfire to return. Individual choices escalate to remake social settings. The local becomes global.

. . .

In most fires what attracts attention is the flame. It occupies the same ground as observers; its dimensions align with a human sense of scale. But the much vaster part of a fire is its plume, which can tower beyond the tallest mountain. What most captured public imagination in the 2019–20 global outbreak of fires were images of great convective columns, slowly spiraling like a black hole in the atmosphere, and palls that turned noon into midnight and drove residents of major metropolises, from Sydney to San Francisco, indoors for weeks. Not since the howling dust storms of the 1930s had humanity's unhappy collision with nature had so dramatic a visual emblem.

The plumes called forth another, perhaps more sinister, symbol. Intense fires can create thunderheads, what are called pyrocumulus clouds. Water vapor rises, cools, and in the right atmosphere can billow into thunderstorms, complete with crystal caps, rain, violent downdrafts during collapse, and lightning. The outwash of the downdrafts can in turn drive fires below the thunderstorm, and the lightning can spark more fires ahead of the winds. The phenomenon is not common—most fires don't generate the energy required, and big fires typically ride strong winds that can bend or break the convective column—but neither is it rare. It's a normal phenomenon that has, in recent decades, become more prominent, frequent, and studied.

And it might stand as a metaphor for how third-fire is interacting with the atmosphere to make more fire in living landscapes. It has always been true that fire feeds on fire, that fire creates conditions that support more fire. Fires in living landscapes have containment baffles, however, that prevent them from overrunning everything. Fires from lithic landscapes don't. Instead they realign fuels, they change climate, they kindle fires ahead of those below the plume. What began as another interglacial from the Pleistocene has, thanks to humanity's fire power, boiled over, punched through old constraints even in the atmosphere, and blown its influence far beyond the flaming front. An epoch of fires has become a fire-informed epoch, a full-blown fire age.

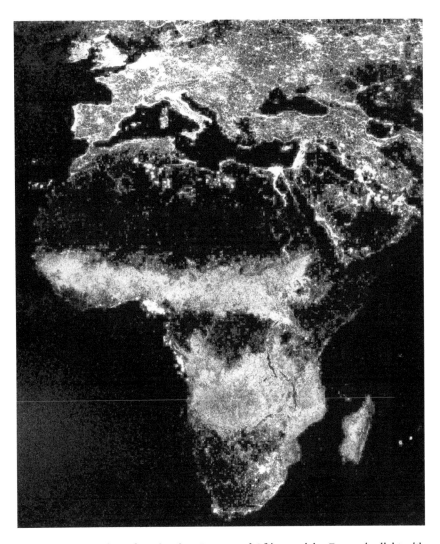

PLATE 1. Two realms of combustion: Europe and Africa at night. Europe is alight with electricity, most of it from combustion, and the rest still part of a fossil fuel society. Sub-Saharan Africa burns living landscapes, except around major urban complexes and offshore drilling sites. Courtesy of Chris Elvidge, NOAA National Geophysical Data Center.

PLATES 2 AND 3. Two fires: lithic and living landscapes at Tallgrass Prairie Preserve in Oklahoma. (a) Bison on freshly greened postburn pasture, with newly burned and still unburned strips in the background. (b) Third-fire, here in the form of an oil well pumpjack, frames even natural landscapes like the preserve, where routine burning and free-ranging bison shape the living landscape. Photos by the author.

PLATE 4. Two visions of fire: Biosphere II in Arizona in the foreground, where all open flame is prohibited, and the 2003 Aspen fire rushing over the Santa Catalina Mountains, where fire is both inevitable and essential. Finding a middle landscape, and a narrative to go with it, is a critical task for the future. Photo by Francisco Medina.

PLATE 5. A future of novel ecosystems, fire and ice. Here two icons of climate change, the polar bear and fire, converge at Churchill, Canada. Photo by Jack Fields. Permission from Science Photos Library.

PLATE 6. Second-fire and third-fire meet. In 2009 power lines from the Hazelwood Power Station started fires responsible for killing 160 of the 173 dead in the Black Saturday bush-fires that blasted over the Latrobe Valley in Victoria, Australia. In 2014 an arson fire in the bush reciprocated by sprinting into the vast open-pit coal mine that supplied Hazelwood. Photo by Mike Keating. Permission from NewsPix.

PLATE 7. The Great Basin burns. Fire, cheatgrass, and climate are acting synergistically to replace the natural sagebrush steppe with an invasive pyrophyte. Photo by Mike Pellant.

PLATE 8. Where the wild and the built collide. America's wildlands are largely an artifact of a fossil fuel society; so are its exurbs, which are a kind of gentrification of former rural landscapes. Two realms of combustion came into conflict at the 2007 Grass Valley fire in California. Photo courtesy of Jack Cohen, US Forest Service.

PLATE 9. Where the wild and the built collide, again. A modern town, Fort McMurray, Alberta, created to mine oil sands but sited in the boreal bush, faces the flames of the Horse River fire, 2016. Residents flee in their cars. Photo by R. D. Darren, Wikimedia Commons, https://commons.wikimedia.org/wiki/File:Landscape_view_of_wildfire_near_Highway_63_in_south_Fort_McMurray_(cropped).jpg.

5 *The Pyrocene*

Take four fires.

On May Day of 2016, a fire of indeterminate origin (though most likely human caused) started southwest of Fort McMurray in Alberta, Canada. Two days later it boiled over into a pyrocumulus cloud, blew over the Athabasca River, blasted into the town, forced a massive evacuation, destroyed some 2,400 structures, and caused the regional economy based on oil sands production to shut down. The fire continued to spread eastward, eventually crossing into Saskatchewan and burning into organic soils, which allowed it to overwinter beneath the snows; its smoke splayed southward to cover over half of the United States. It was not declared out until August 2, 2017. Its final size was reckoned at over 1.2 million acres. Its final costs, direct and indirect, were estimated at $9.9 billion.[1]

The Horse River fire, as it was officially named, was not the largest in Canada's recorded history nor the first to overrun communities. The settlement era had witnessed towns like Porcupine, Cochrane, and Haileybury destroyed; and in 2011 a similar fire, under comparable circumstances, had stormed out of the surrounding boreal forest to burn into the urban core of Slave Lake, also in Alberta. Boreal forests burn, and communities nestled within the forests, if not shielded, will burn with them. The Horse River fire was historic, but not unique. Watching the flames return was in some ways like watching a plague long thought banished return.

What makes the fire particularly interesting is its relationship to the two realms of combustion that were informing the region's pyrogeography. Modern Fort McMurray exists to service the Athabascan oil sands operations just to its north. Here, amid some of the dirtiest of fossil fuel production, embedded within some of the most flammable forest on Earth, oil is extracted from lithic landscapes. But as oil is squeezed out of stone by heating, so Fort McMurray is squeezed between living and lithic landscapes. In May 2016 those two realms of fire, usually segregated, met. That so many of the images showed flames soaring through the woods while the foreground filled with vehicles—cars that had melted, cars carrying fleeing residents, trucks hauling gasoline to power those cars—perfectly captured the collision. The two burning ends of humanity's combustion candle merged.

· · ·

A year later, June 17–18, 2017, the fires came to central Portugal. The weather provided a classic formula for big burns—hot and dry, from a historic heat wave; windy from Tropical Storm Lucifer. Dry lightning and human ignitions sparked 156 reported fires, mostly southeast of Coimbra. But ignition requires kindling, and it was the extensive presence of extraordinary fuels that powered small fires into a handful of conflagrations. The flames fed on abandoned fields and on forest plantations of exotic pines and eucalypts. The blowups reached 1.3 million acres, overpowered firefighting operations, destroyed communications networks, broke power lines, and killed at least sixty-six people, most of them trying to flee flames around Pedrógão Grande. Another outbreak came on October 15-17. With 2 percent of Europe's landmass, Portugal managed to account for some 60 percent of Europe's total burned area for the year.[2]

The fires were not, as sympathetic critics noted, unprecedented or unexpected. Portugal has a Mediterranean climate that favors an active fire scene, though one that Portugal has kept in check over centuries through close (even overintensive) cultivation. By the mid-1970s with the end of dictatorship, Portugal's admission to the European Union and entry into a modern economy, which rendered traditional agriculture unattractive and inadequate, had inspired a wholesale exodus from Portugal's rural lands to its metropolitan areas. The countryside overgrew with flammable shrubs and plantations of fire-thirsty woods. The traditional infrastructure to contain burning faded. Much as exurbanites reoccupied some villages, so modern fire agencies assumed responsibility for fire protection.

It was not enough—could never be enough except at extreme costs. Since 1974 the number of fires had risen sharply, the burned area had grown (occasionally, explosively), and fire suppression had failed during extreme events—precisely the occasions it was most needed. In 1991 the region had experienced major fires. In 2003 fires burned around and even into Coimbra; they came again in 2005. Perhaps no country in the world had suffered proportionately with so many bad fires. Still, it was a regional problem felt in Spain, southern France, and particularly Greece. In fact, fires burned in Provence at the same time they flared in Portugal.

Unlike Fort McMurray, villages like Pedrógão Grande were not newly erected amid a flammable landscape, but ancient townsites that had a new landscape grow around them. A fossil fuel economy underwrote the scene, not directly but indirectly, as the countryside was abandoned and the young left rural economies for modern ones like those at Lisbon and Oporto. A firescape that had been organized by traditional agriculture unraveled under the pulls and tears of an industrial economy. If Fort McMurray, in its way, was an outpost of an advancing frontier, the towns of central Portugal were those of a retreating one. The propelling condition was less a global climate than a global economy.

The small field and pasture fires that had once characterized the region yielded to the removal of second-fire and its replacement by a macabre fugue between first-fire and third. Lightning kindled most of the fires, internal combustion tried to counter them. The rise in feral fire tracked the loss of second-fire. On the N-236 highway outside Pedrógão Grande, where the two realms crossed, thirty people perished in their cars and another seventeen near their vehicles. Here were the Pyrocene's lines and fields of fire.

. • .

Once again, there were foreshadowings. The town of Paradise, California, had had warnings before the Camp fire rampaged through the town—was long recognized as a hazard, had experienced thirteen large wildfires nearby over the past 20 years, had a nonlethal dress rehearsal with two outbreaks in 2008, and in the summer of 2018 had benefited from a regional grant to support mitigation measures. At 6:15 a.m., an hour before sunrise on November 8, 2018, with an east wind howling, a faulty power line threw sparks. Within 18 minutes the fire was 10 acres; an hour and 45 minutes later, its embers showered Paradise, and within another hour, the town was a wreck of smoking ruins, concrete pads, and melted autos. Not until rains fell 17 days later was the fire finally extinguished.

Losses were extensive. The fire killed eighty-five people, destroyed 18,661 structures, disrupted the communications infrastructure, deluged the Bay Area and surrounding Central Valley with record air pollution, and displaced thousands of residents—yet another cavalcade of fire refugees. Such are the synthetic materials of modern life that the town was declared uninhabitable until the toxic debris, now freed from containment, could be properly cleared (the cleanup contract ran to $3 billion). Insurance claims strained particularly small companies. Pacific Gas and Electric, the utility whose grid had failed, faced claims for damages as high as $30 billion, and filed for bankruptcy protection while CEOs faced charges for criminal negligence. To avoid repeats, utilities began turning off power during red flag days; this affected millions of customers. Insurance companies raised rates punitively. It was the costliest global disaster of the year. The knock-on effects will continue for many years.[3]

California is built to burn and occasionally to burn explosively—fires are hardwired into its physical geography. At the same time the Camp fire slammed into Paradise, Southern California was grappling with the Woolsey fire, the largest in the state's recorded history. In the 12 months preceding the Camp fire, California had experienced two waves of fire busts. An autumn 2017 outbreak had affected both Northern and Southern California; one of that swarm, the Tubbs fire, swept into Santa Rosa and incinerated 5,643 structures. In August 2018 a bust of big fires broke out in Northern California, including one that tore into Redding as a fire tornado. California and fire were no strangers.

But California's historic pattern had been one of a big fire or fire bust followed by a quiet period of perhaps 5 to a dozen years. Here were three big blowouts in a year (followed by others in 2019 and 2020). Conflagrations were not unusual; serial fire sieges were. Something had changed, and observers looking for accelerants pointed to a perturbed climate, to the legacy of land use that had scrubbed away small fires and left only big ones, to a disbelief that urban fires could move from the exurban fringe to (or over) the urban core, and to a misplaced confidence that more engines and air tankers could hold the line. California had a fire protection infrastructure unrivaled in the world; it had the five largest fire departments in the country and a century of aggressive suppression experience. By 2020 it could no longer pretend that a fire agency could stand between California's firescape and a preferred way of life that chose to ignore fire. California had crossed a threshold.

· · ·

So, it seemed, had Australia. The 2019–20 bushfire season started a month early, in September, and burned without cessation into mid-February. A prolonged drought; record heat; bouts of dry lightning; human ignitions, from helicopter landing lights to beekeeping operations; a firefighting organization, mostly staffed by volunteers, stretched to its limits—even for a continent renowned for its bushfires, this was an outlier season. Many observers considered it a signpost to the future.[4]

Preliminary reports sketch the contours of the raw narrative. The fires visited every Australian state and territory, though the most severe burns ranged from Kangaroo Island to the coast of New South Wales, an estimated 27.2 million acres. With almost malevolent cunning, the worst fires concentrated on the protected forest lands and national parks along the Great Dividing Range and Gippsland. At least thirty-three people died, four firefighters among them. Residents in towns like Mallacoota fled to the beach, where they were evacuated by the Australian Navy. By the start of February the Australia Institute estimated that 57 percent of Australia's population had been directly affected by the fires through flame or smoke. Smoke palls blanketed Sydney, Melbourne, and Canberra for days, then through weeks; citing unhealthy air, Australia Post halted mail deliveries in Canberra. The smoke affected vineyards spared from flames. It slowed economic activity, even—with delicious irony—forcing BHP to halt coal mining for a while. Tourism suffered, both domestic and international. Municipal watersheds were impaired; estimates of dead fauna soared, to as many as a billion; ecologists worried about the ability of even Australia's fabled, fire-tolerant flora to recover, especially amid a changing climate. Suppression costs swelled, particularly once officials decided to pay volunteer fire brigades who were called on for extended duty that went well beyond norms. The federal government committed $2 billion to recovery efforts, though estimates of the total damages ranged as high as $100 billion. A royal commission was planned. A full roster of impacts will likely take years to tally.

No question—Australia is a fire continent, conditioned by millions of years to drought, pyrophilia, and a capacity for conflagration. Imagine California the size of the contiguous United States, but drier, with a quarter of its land mass subject to the Asian monsoons, and a record of human fire use dating back at least 50,000 years. Through two centuries of European colonization, historic conflagrations had filled up the calendar with a Red Tuesday, an Ash Wednesday, and black everything else. These were not just environmental but political events that had inspired inquiries and royal commissions. But the Black Summer of "forever fires" felt different. The tempo of bad burns had quickened, and their damages were worsening. Climate change and varieties of controlled burning became talking points—flash points amid a divisive politics. Pyromancers stared into the flames and decided the space-time continuum had torn a hole in southeastern Australia, a portal to the Pyrocene, through which one could see the dimensions of a maturing fire age.

. . .

Now, look beyond the four flare-ups.

It was possible to argue that these flash points occurred in places long subject to fires and that the global constellations of flame tracked by satellites were so many sparkles of light from the ecological equivalent of exploding fireworks, that they were media distractions from the real workings of the planet. Certainly fire was exploited to animate other messages. The frontiers of fire gnawing into Amazonia and Kalimantan were not primarily fire problems or indicators of climate change, but issues of politics and indices of a global economy with fire as an enabler and climate an enhancer. The smoke palls of northern India mapped where petrol-powered pumps for irrigation met traditional burning of fallow and stubble. The speckled band of flame that ranged across the boreal forests from Siberia to Alaska— "fires in the Arctic," as labels went—burned in landscapes primordially prone to crown fires. Places like the southeastern coastal plain of the United States practiced prescribed fire at a significant scale; Florida burned 2.5 million acres a year. The northern savannas of Australia were shifting fire regimes at a subcontinental scale from large, late-season burns to patchier, early-season burns. Much of what the world witnessed was a sharpening and deepening of legacy firescapes and a media magnification of eye-grabbing memes.

But the big fires had bulked up, and fires were also moving into places that had long contained them in agricultural settings or that had not known wildfire for decades and now experienced feral fires. These were places like Bastrop County, Texas, and Gatlinburg, Tennessee; like rural Britain where moors burned from the Scottish Highlands to East Sussex; like Sweden, Germany, and Italy where fires broke out amidst record-shattering heat waves and dry lightning. The occasional flare-up was an anecdote; regional outbreaks were a statistic; repeated blowouts were a narrative.

The flip side was, paradoxically, those places that should have burned and didn't. Earth's fire story was not just about the visible, the sudden, and the novel: the invisible, the incremental, and the traditional were equally part of the emerging order. The absence of fire where it should be was as critical, if less conspicuous, as its exaggerated presence. While the economics of missing fires was generally unclear, not least because economists relegated environmental matters to externalities, the ecological effects became more insistent each year, not only in terms of biodiversity, but in biological goods and services, watersheds, and the potential fallout for future fire. Typically, fire exclusion rendered sites more prone to more intense burns with more severe consequences. Like early steam engines used to drain coal mines, fossil-fueled combustion was channeled into land clearing that then fed abusive burning.

This was a pathology of the developed world. In places that lacked wildlands, fire's absence manifested itself through a deterioration or loss of cultural landscapes—heaths in northern Europe, and rough pasture in Mediterranean Europe, for example. Such places recorded the collateral damages imposed by a fossil fuel society. A deficit of good fire was less telegenic and dramatic than a surplus of bad fire, but their environmental payouts could be equal.

Nearly everywhere, it seemed, was connected to the inextinguishable presence or malign absence of fire. The scatter diagram of fire points began to cluster and align along new industrial lines and fields of fire. The separate scenes became linked, as human societies and economies were joined into networks by fossil fuel transport. Satellite imagery revealed a global pyrogeography of hot spots. The ramifications of fossil-fuel-driven climate change suggested a more universal bond. What nearly every place shared was that fire was always present, as cause, consequence, or catalyst, as the unalterable companion of the human hand and mind. A new world order on fire acquired coherence and mass.

Draw a regression line through that scatter diagram to make a narrative, and then connect the dots into an image. The narrative arcs through the long history of humanity and fire. The image is the fire analogue of an ice age. Together they make the Pyrocene.

Fire Age

What does a full-blown fire age look like?

Here an extended analogy to the Pleistocene's ice ages provides a rough template. Some effects are direct, some indirect; the role ice assumes in the one, fire takes in the other. In full-blown ice ages there were huge biogeographic shifts, an expansion in the dominion of ice, drops in sea level, mass extinctions, all amid a radical reconstitution of climate as ice made a world more favorable to ice. Amid all this the hominins appeared and then thinned to a single species. In a fire age we can expect big biogeographic shifts, an expansion in the dominion of fire, rises in sea level, mass extinctions, all amid a radical reconstitution of climate as fire makes a world more favorable to fire. In truth, climate history has become a subnarrative of fire history. The surviving hominin, the sapiens, may well change its genomic nature or spin into extinction.

All analogies collapse eventually, and some can quickly veer into the absurd. Ice is a substance, fire a reaction. Ice is a single mineral in a single state more or less oblivious to context. Fire is a notorious shape-shifter that integrates everything around it. Ice moves at a tempo of decades and millennia; fire, as quickly as a gust of wind. Ice is nature as modernist, a single vision ordering a world; fire, nature as postmodernist, all about context. Yet in their capacity to inform landscapes, they display comparable ranges and power. One can map with eerie fidelity onto the other. The concept of a fire age can do for understanding the Pyrocene what the concept of an ice age has done for the Pleistocene, and while the idea may be new, the process has been underway since the onset of the current interglacial. With almost surgical cunning, fire has been systematically driving off ice along with the habitats and creatures it inspired.

Places that already have fire will see more of it, or watch its regime change toward larger, more frequent, and more eruptive outbreaks. Wetter grasslands and savannas, such as tallgrass prairie, longleaf pine woodlands, sourveld, and cerrado; boreal forest and muskeg; Mediterranean shrublands and fynbos; pine, oak and hickory, and acacia and miombo woodlands subject to frequent surface fires—these are the equivalents of ice sheets. All will continue and perhaps not simply experience fire as part of a suite of stresses, but watch fire become their informing disturbance. Places that have fire episodically, like peat, heath, moor, and organic soils, may see it become more formative as a fire-climate alliance pushes fire regimes outside their historic ranges and fire-prone biomes become fire-dominant pyromes.

Fire-intolerant places may flip into fire-tolerant ones. Land clearing and logging, with fire as a catalyst, can convert rainforest into pasture and palm oil plantations. Once burned, a land can be more receptive to further fires, and repeated burning can prevent recovery to former conditions. The land toggles into other arrangements. Likewise, invasive grasses such as cheatgrass, buffel grass, cogongrass, and gamba grass, can work with fire, like knives fashioned into scissors, to cut through existing landscapes and leave a more fire-thirsty biome; some 60 million acres in North America are infested with cheatgrass, most converted from sage steppe, with much more to come. What in evolutionary time had occurred with the appearance of C_4 grasses is now occurring with the human-assisted transport of exotic pyrophytes to transmute incombustible biotic dross into flame-receptive fuel.

The fires become a self-reinforcing process, with the more (and earlier) flammable cheat driving out competitors. Invasives (like people) thrive on disturbance, putting them and an inextinguishable source of ignition in close proximity. They cling to fringe settlements. They follow roads into logged or cleared areas, acting as fuses to carry fire to new sites, cracking open once fire-intolerant ecosystems and exposing them to fire infections. The scale can be subcontinental—Amazonia, Kalimantan, the Great Basin. These might be likened to the pluvial lakes that flooded immense regions outside the ice sheets proper.

The organic soils frozen into permafrost have their analogue in the organic peat now exposed to fire in the tropics and the boreal zones. The greenhouse gas release from peat burning in Indonesia is estimated at 10–40 percent of the annual greenhouse release from fossil fuels; in some years it is the largest single contributor to CO_2 globally. As the organic-rich permafrost itself is exposed and melted, yet more biomass—once in cold storage—will become available as fuel, release greenhouse gases at a continental scale, and feed back into the system. These are instances not of living landscapes burning and then recapturing their carbon and nutrients in new growth, but of a prolonged, geologically measurable transfer from biomass untouchable by flame to combustible biomass accessible to it. The process by which the Pleistocene had stockpiled carbon, and so further cooled, the Pyrocene is reversing, liberating those reserves by fire and so further warming.[5]

The outwash plains of sand and silt that winds carried from the ice's edges have their counterpart in the smoke palls from far-ranging, long-lingering fires. Because fire, unlike ice, is not a substance, it does not remain permanently on the land; instead it brands biomes, then reappears. Similarly, its by-products are not materials abraded and blown into geomorphic caches, but ephemeral effluents. Smoky days have mutated from seasonal nuisances to the public health crises of megapalls. Metropolises well distant from flame can suffer the consequences of inextinguishable burning. In August 2020, public health officials in Denver advised residents to consider constructing "safe rooms" to avoid the toxic cocktail that wildfire smoke, from California as well as Colorado, and industrial emissions had stirred together.[6]

Those modern cities might themselves have their analogue in mountain glaciers—those ice caps and more localized ice clusters not sprawled at continental dimensions. Modern metropolises are fire places, organized around the power grids of industrial combustion and the ICE (internal combustion engine) of third-fire, reshaping their landscapes as ruthlessly as water ice. Fire of living landscapes has been banished as fully as possible, but such places are as completely organized around lithic landscapes as the Sierra Nevada of California or the Alps were around ice.

For the equivalents of periglacial effects, look to the ecological knock-ons produced by a warming climate. Diseases and their vectors. Insect outbreaks. The breakdown of seasonal barriers that had previously confined irruptions to patches but now allow ripple effects to magnify into the ecological equivalent of rogue waves, like mountain pine beetle sweeping from British Columbia to Colorado, a rolling thunder of biotic disruption that rearranged fuels as it went. Fire season lengthened; in the American West this manifested as earlier, drier springs, which then interacted with human ignitions. Fires burst into modern communities—patches stuffed with synthetics and heavy metals—and did here what they did in living landscapes; they broke down inert masses and liberated chemicals into the wider surroundings. Instead of renewing, these burns poisoned. The economic shocks ramified; the 2017 and 2018 fires in California drove the largest public utility in the world's fifth largest economy into bankruptcy, while insurance companies reconsidered what they were willing to underwrite. With ice, collateral consequences were primarily physical. With fire they are primarily biological and social.

And then there are deserts, some of which will expand and some of which will emerge from grasslands, shrublands, and steppes as a reorganizing climate breaks down the wetting and drying that had underwritten their flora and fires, much as an ice-favoring climate had done in the Pleistocene. Not every place has to burn to be influenced by fire's reach. It's enough for combustion's consequences, in this case on climate, to shape biogeography.

Nudges, when repeated, can push species and biomes across a threshold. More severe fires will spark greater mortality by burning longer and hotter and especially by razing nature sanctuaries. The mottled nature of most fires—patchy in space and time—will be lost, and with it traditional refugia, and repeated burns can hinder the habitat from recovering, and populations of wildlife from rebuilding. The frost-thaw bellows of the Pleistocene had created a fifth extinction. The fire-catalyzed Pyrocene is sparking a sixth. The lost fauna will affect flora, which is to say fuels. By no longer consuming browse, the missing megafauna of the Pleistocene freed more combustibles in fire-prone lands. Contemporary extinctions will shape fire as well—how, is little understood.

Some landscapes will be immune to fire, as some were to ice. Deserts were too arid to support ice or lakes, and they are too barren to carry fire, save when the odd rains temporarily grow fuels. Nor will oceans burn, as the Arctic and Antarctic had previously plated over with pack ice. Instead, they will feel the impact of the fire age through oil spills and the indirect consequences of global warming. The ice will melt. The seas will rise and their acidity increase. The biology of the oceans will reorganize as currents reroute, coral reefs dissolve, and continental shelves expand. Thanks to its capacity to reconstitute the Earth's climate, every place touched by the atmosphere will be touched by fire.

The point is not that every place will burn or burn more, but that humanity's fire practices and habits will influence more and more of everywhere. Anthropogenic fire has always been a means to force, to leverage, to sculpt, and to explain, but at particular sites. In the globalizing processes of the Pyrocene, its reach will be planetary. Instead of implacable ice we will have unquenchable fire.

.   .   .

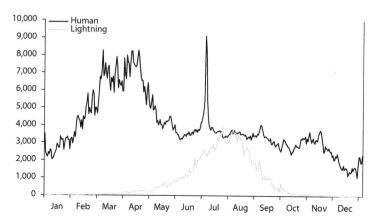

FIGURE 4. Ignitions by humans (darker line) and lightning (lighter line) in the United States by month. Though an industrial society, the United States displays the basic template of annual burning characteristic of most of the world. Note the overwhelming proportion of anthropogenic ignitions and how the two sources, lightning and human, occupy separate seasons. The creation of public wildlands dampens the number of human ignitions, since such lands typically prohibit anthropogenic burning, except by designated agents. By interacting with the opportunities created by changing climate and land use, humans are expanding the fire season. Data source: US Forest Service, courtesy of Jennifer K. Balch and Adam Mahood, www.fs.usda.gov/rds/archive/catalog/RDS-2013-0009.4.

And the hominins? In the Pleistocene they had to adapt to the micro- and macroworlds created by ice. In the Pyrocene they have to adapt to the worlds their fire practices have created. The sapiens have become not just the dominant hominin, but the dominant species on Earth. They could not kindle ice, tend it, herd it, extinguish it, process their food with it, or use it to extract usable goods from rock, trees, or water. Ice acted, humans reacted. But fire they can start and stop—can manipulate to remake foods, eco-systems, and the world around them. Their ignitions can interact with climatic shifts to extend fire seasons, to redefine fire regimes, to carry fire where it would not naturally exist.[7]

Yet while humans can manipulate fire—can refashion Earthly fire into their own, have remade fire in their own image—they cannot control all the consequences. They can disrupt more easily than they can direct. Their ignitions can interact with climate change to lengthen fire seasons; can interact with invasive grasses to overturn biomes; can redistribute through plumes toxic metals and radioactive soils. Like individual drugs that unexpectedly create new outcomes when combined with other drugs, fire's secondary effects are interplaying with one another with consequences far beyond the reach of the flames. More than just synergy, it's a matter of scale. Humanity's combustion habits have measurably affected the atmosphere, the hydrosphere, the lithosphere; they have interfered with biogeochemical cycles; they have rewired the energy flows of the planet. Everywhere they are driving into extinction the relic species, landforms, and ices of the Pleistocene. They are making the Earth uninhabitable not only for other creatures, but for themselves. What began long ago as a mutual assistance pact between hominins and fire looks more and more like a Faustian bargain.

Living with Fire: The Principle

For decades the American wildland fire community has, with near unanimity, repeated the message that we must learn to "live with fire." Originally this meant that attempts to extirpate fire from every landscape were misguided, that fire was both inevitable and necessary, that instead of mindless attempts at suppression, we had to renegotiate our relationship to fire. They had in mind mostly wildlands. But as the planetary sweep of the Pyrocene becomes apparent, it applies to all of humanity's fire practices.

What does that mean? It means we need to understand that fire is not going away and that, in many of its expressions, we can't afford for it to leave. It means relocating fire intellectually from wild fringes to a central presence. It means that industrial fire's ecology matters as much as the fire ecology in living landscapes. It means our industrial transformation did not banish fire; it only displaced it, stuffed it in machines, and replaced tamed fire in many landscapes with feral fire. It means we can take flame out of our built environment, but not out of countrysides and wildlands. It means rebalancing the three fires, specifically ratcheting down third-fire and ratcheting up second-fire. It means working with fire rather than against it. It means making fire once again our best friend instead of our worst enemy.

We don't need new science or more science; we already know what needs to happen (in truth, we used to know much of it before we got greedy and forgot). We know we need to replace fossil fuels with another source of energy. We know how to protect communities and critical assets like watersheds from bad fire. We know we need to get good fire back into landscapes, not with the engineered precision that drives pistons, but within the range of tolerance of most biotas—the sloppiness and variability of fire is, in fact, an asset that builds resilience. We know we need lots of controlled fires from lots of sources, all with their peculiarities: there are many varieties of prescribed burning, and they all have their place, often the same place. All this is simple enough to say, and it disguises the thousands of decisions that go into each choice. But then we don't have to parse every fragment and factor, because fire will do that work for us. It synthesizes: it will integrate better than the most powerful supercomputer if we allow it. We have to learn from experience with real fires, not through computer-generated simulacra.[8]

Even the most ambitious revolution will be slow and incomplete. Many current conditions are baked into climate, seas, land, and terrestrial life. We face a long, painful hangover from our overindulgence of fossil fuels. If we abandoned fossil fuels tomorrow, it would take decades, perhaps centuries, depending on how vigorously we exert ourselves, to purge excess greenhouse gases from the atmosphere. A changed climate will lag for a worryingly long time. Similarly, even if we substitute renewables to power our vehicles and make fire surrogates for fertilizer and biocides, they will not by themselves change how we live on the land, only the source of the power behind that way of life. We will still have a landscape structured by the era of fossil fuels: it will have the same form, only infused by a different power source. Cars will have batteries rather than fuel tanks; but we will still be driving cars and using asphalt roads, and we will have exurbs at risk from wildfire. We will be diminishing the output of greenhouse gases, but not inhabiting a different world that better accommodates fire around us.

A stable climate, an unstable land usage—this was the formula for fire crises in the industrial world before a fire-perturbed climate globalized the scene. The organization of that world will persist. Abolishing industrial combustion won't make bad fire go away, nor ensure good fire will happen. We will have to manage fire in ways that, for developed nations, have not been seen as necessary for a century or more.

Earth's fire scene holds not one fire, but three, and of all the unanticipated outcomes that will result, how these fires interact among themselves amid a churn of ecological pathologies and a cascade of disruptions will define the fire ecology of the future.

First-fire and second-fire, first-fire and third-fire, second-fire and third-fire—at various times and places they compete, complement, and collude. But each integrates its surroundings, and as global change thickens, with not only climate but the whole of the planet subject to upheaval, each fire will respond; and then they will respond to the presence and postburn landscapes of one another. It's likely that the fires will exceed not only our capabilities to control them but our ability to forecast what kinds of control will work.

A few projections do seem secure. Because we construct it, the built landscape is under our relative dominion. It was once common for cities, fashioned out of the same materials as their countryside, to burn in much the same rhythms. That stopped when industrial societies turned to less-flammable materials and more-stringent codes and zoning. For a while exurbs, which flourished beyond metropolitan boundaries, also seemed outside those regulations. That impression was an illusion. Recreating the old conditions recreated the old fires. The way to shield towns from flames is by rediscovering the same methods that stopped urban fires in the past. The problem was misdefined as primarily about wildlands when it was actually about urban enclaves. Such places are not houses in wildlands so much as towns with peculiar landscaping. Modern cities don't want open flame; many wildlands do.

The tougher issues will involve the countryside. Here the way to cope with our coming fire age is, paradoxically, with more fire, but a shift among fires back to second-fire from third. Meanwhile, expect strange alloys, some of which will yield bad fires: the power lines that arc during high winds and spark wildfires that crash into towns like Paradise; the traditional burning of stubble in the Punjab that, stirred into a common cauldron with the emissions of industrial combustion, smothers New Delhi in a stew of smog; restored natural fire in Rocky Mountain wilderness that can blanket valley towns like Hamilton and Missoula with smoke. What was a benign burn in the past may now be a problem fire. Other alloys, in contrast, will seek to turn problem combustion into good fire: The West Arnhem Land Fire Abatement Project in northern Australia is trying to reconcile all three fires. The traditional burning regime of Indigenous Australians—small patches early in the season—is replacing large, sweeping burns at the end of the season. The process sequesters more carbon in more trees, and in a brilliant twist, it uses carbon credit funds from BHP's offshore natural gas project in the Arafura Sea to finance the work. Third-fire money supports second-fire restoration that stores more carbon and enhances ecological goods.[9]

Expect more strange amalgams: controlled fire to reduce fuels that threaten towns and critical ecosystems; small fires to protect carbon stocks in planted and thickening woods from big fires; good fires to stabilize biomes, to substitute manageable smoke for enshrouding palls; and catalytic fire to accompany other interventions—fire as a driverless car with its enormous capacity to integrate makes available many points of access and intervention. Current fire knowledge means that we can anticipate only a fraction of the surprise interactions that are sure to result.

There are three kinds of fire—natural, anthropogenic, industrial. There are two realms of combustion—one in living landscapes and one in lithic ones. The task for the future is to divide three into two and have something left over.

Living with Fire: Practices

A fire age happens globally. Fires occur locally. There are hundreds of words, for example, for slash-and-burn agriculture, all specific to particular places and peoples, although we can lump them together under the label swidden. In the same way, there are hundreds of tactical practices that characterize our relationship to fire on the land. But we can abstract them into four strategic groupings: we can leave fire to nature, we can substitute controlled fires for wild fire, we can alter the environment under which fire of any sort burns, and we can prevent fires from starting and suppress those that occur. Which is right? All of them, and often all of them together in various proportions.

Leave to Nature

Nature has more pathways than humans have tools, and more information coded into genomes and biomes than people have ideas. Allowing nature to work itself through changing fire regimes makes sense; nature has been at it for eons, has accommodated fire through primordial icehouse and greenhouse climates, and has survived at least five global extinctions. That record should give pause to humanity's ambitions. If humans cease their insults, nature will find ways to heal.

As with the wild generally, people cope best when wild and human habitats are separate. This is the situation, for example, across much of the circumpolar boreal environment, where economics, ecology, and firefighting capacity argue to let nature's fires do nature's work. Traditionally, Canada has chosen not to suppress fires beyond a line of control at roughly 60° latitude; much of the Russian taiga lies outside commercial logging and mining and is left to burn, or if an initial attack fails, the fire is permitted to linger since it is effectively beyond control. The borders between wild and inhabited are blurring, however, and the stakes are rising since global warming is felt more acutely in the higher latitudes, and the liberation of immense reserves of carbon stored in organic soils and permafrost could unleash a blowout of greenhouse gases if freed by fire. How to manage the transition to a new regime will likely involve complicated negotiations between what to leave and where, if even possible, to intervene.

Alaska has refined the concept and turned it to ecological advantage. The Alaska Statehood Act (1959), Alaska Native Claims Settlement Act (1971), and Alaska National Interest Lands Conservation Act (1980) reapportioned land ownership, and hence fire responsibility. In addition, it has a small population in a vast, fire-prone landscape, much of it committed to wildlife refuges and legal wilderness. An interagency program crafted a suite of novel fire plans to replace an unworkable policy of suppression. Outside cities and villages, the strategy allows fires to roam, or it loose-herds them by burning along natural barriers like rivers. The rationale is based not just on calculations of cost and capability, but also on the ecological benefits of burning in a fire-dependent forest. The upshot is a close approximation to what a natural regime would be. Something like this model is applicable wherever there are lightly inhabited outbacks.

There may well be calls to limit the burning, so vast is the boreal biota and so worrisome is the prospect of well-stored soil carbon in muskeg and permafrost being loosed to an already carbon-marinated atmosphere. Some sort of patch burning within the boreal forest, including prescribed crown fires, may be a consideration. Or it may be that the resources humanity can commit to management will have to be concentrated at sites where fire poses a more immediate threat to human settlements. In a huge place with fewer options, humanity may choose what it prefers, with nature deciding the details of what follows.

The strategy is also useful in protected nature reserves, especially when the objective of the land is to preserve natural conditions and processes as much as possible. In the high valleys in California's Yosemite and Sequoia and Kings Canyon National Parks, in wilderness sites in the Mogollon Mountains of New Mexico, and in the Selway-Bitterroot Wilderness in the northern Rockies, a good fraction of lightning fires were allowed to run their natural course beginning in the 1970s. The cumulative consequences, the layering of new burns on old ones, have yielded fundamental knowledge about how fire worked before suppression programs intervened, but they have also made a landscape more robust than surrounding lands that have not had the tempering effects of repeated burns. These are prime sites, however, for which restoration can rally social sentiment and rely on cultural capital, probably not a strategy that can scale across landscapes of mixed use and urban settlement.[10]

Substitute Tame Fire for Wild Fire

If fire there will be—must be—then replace fires of chance with fires of choice. Substitute some variety of planned and controlled fire for fires started by lightning, accident, or arson. This is an ancient art, dating back before the sapients. What we did by cooking food over a hearth, instead of scavenging among burns, we did by burning landscapes to encourage a more favorable habitat and one that was less likely to burn in unfavorable or dangerous ways.

This is, in fact, the premise behind second-fire, and it has been one of the casualties of third-fire. It has returned to prominence because it had to—because in the absence of routine burning, many fires became utterly uncontrollable and damaging, because ecological integrity unraveled, biomes rotted, and environs became unusable without it. In the second half of the twentieth century, the restoration of fire by deliberate burning—prescribed fire—lay at the core of American thinking about fire in most public and many private lands.

When he coined the term *prescribed burning* in 1942, Raymond Conarro envisioned the practice as a compromise between feckless folk burning and mindless suppression. It was conceived to be scientifically informed and bureaucratically disciplined in ways that left, or seemed to leave, its practitioners in control. The practice flourished first and best in the longleaf pine savannas of the southeast coastal plain. Florida, in particular, had long resisted efforts to impose fire exclusion, and agency after agency traded that fantasy for a doctrine of routine burning. The US Fish and Wildlife Service adopted a variant of prescribed burning in the 1930s, the US Forest Service in the 1940s, and the National Park Service and Florida Forest Service in the 1950s. By the 1960s, promoted by Tall Timbers Research Station, it became the basis for agency-wide policy reforms by the National Park Service, and by the 1970s, for the Forest Service. In an effort to incorporate natural fire in wilderness and backcountry areas, advocates advanced the concept of a prescribed natural fire. By 1978 fire by prescription became the basis for national policy.

The prescribed fire is a set piece in which date, time, place, conditions, and outcomes are spelled out. The prescribed *natural* fire has to rely on lightning's lottery but otherwise mimics the template. In its original avatar, prescribed burning was most common in pine-savanna landscapes or pine plantations and involved cool-dormant-season burns. With experience that model evolved to incorporate hotter fires across other seasons. Most of Florida accepted almost any fire, as many fires as imaginable; the old saying was that Florida burned twice a year. It helped that, thanks to open-range herding, a culture of burning never expired. To reduce concerns over escaped fires, a Prescribed Burning Act was enacted in 1990 that revised liability standards to encourage more burning. The Florida template has spread throughout the Southeast and beyond.

With experience, the Florida model has moved from a relatively uniform practice of strip firing in the dormant season to a more mixed regimen of burning across all seasons and with a wider range of effects. If prescribed burning is to expand, however, it needs to be more nimble, less specific to particular sites and times, more able to roam across larger landscapes and longer burning seasons, more grounded in experience than in simulations—science will supplement, not direct. It will, in brief, more closely emulate historic folk patterns and practices in which fire was routinely present. It may more resemble a process of fire foraging than of fixed set pieces. A changing climate will bring opportunities as well as hazards. Prescribed fire will have to adapt.

The project stalled in the western United States. The reasons are many: The physical geography of the West is different—more mountainous, more prone to explosive winds, more biotically varied. So is its cultural geography—more urban, more land in public administration, more recently settled and without a resident fire culture. The upshot is that fire by prescription has not been able to achieve the scale needed to restore fire in the West. The prescribed natural fire also stumbled, since it was nearly impossible to halt a fire that exceeded its prescription: most simmered but a few blew up, and those escapes resulted in big costs and political messes. Escaped prescribed fires killed people, burned houses, and smoked in communities. In 1988 prescribed natural fires burned over 40 percent of Yellowstone National Park; in 2000 an escaped burn rampaged through Los Alamos, New Mexico, home of the lab that performs most of the country's research on nuclear weapons; in 2012 the Lower North Fork prescribed fire set to improve conditions in Denver's municipal watershed escaped, burned twenty-two houses, and killed three residents. Such prescribed burns gone bad affect prescribed fires everywhere.[11]

The American West will require a different variation—is already experimenting with hybrids that fuse prescribed burning with suppression. This is the concept of the managed wildfire. Fire officers are picking up the other end of the firestick. Instead of seeing every fire as a problem, they wonder if it could be an opportunity, or as some fire management officers put it, a fire is innocent until proven guilty. If a fire potentially threatens high-value assets (like a community), they concentrate suppression there and back off otherwise ("draw a box") to a place where they can burn out, often over extensive areas and extended times. The outcome is part suppression, part prescribed fire done under urgent conditions. It's a case of riding the tiger. There is no pretense that the results will conform to preestablished goals, only the expectation that a significant fraction of what burns will fall within the range tolerated by prescriptions. It's a way of getting good fire on the ground at quantities sufficient to make a difference.

Meanwhile, a more meticulous campaign is underway, most robustly developed in Australia and with a scattering of kindling points throughout the American West, to promote "cultural burning." The ambition is, through recovered fire practices, to restore both land and culture among Indigenous peoples. The suppression of fire practices was part of colonizing the land; restoring fire is seen as a means to recover some of those losses. It's fire as renewer, as a catalyst for a revived heritage as well as a recovered legacy landscape. The problem was not limited to colonial settings: the suppression of traditional knowledge occurred equally within Europe and among settlers as elites denounced fire use as primitive and irrational. In all these instances, restoring a fire culture is seen as critical to larger issues of cultural identity.

What the emergence of new arguments shows is that, in fact, there are many varieties of prescribed burning. Some aim at reducing fuels, some at boosting ecological health, some at restoring traditional or cultural heritage, some at letting nature off a leash, and some at just getting fire on the ground without scaring towns, smoking out tourist havens, and casting dollars to the wind. Prescribed fire is less a vaccine than a booster like flu shots— needed regularly, not completely effective for all recipients or all times, but better than leaving the population to chance. It's ecological maintenance, closer to spring cleaning than heroic intervention, more wellness program than emergency room. However we choose to characterize it, we'll need a lot more good fire of all kinds, a ritual of burning that has no end. It will continue forever.

Change the Character of Fire's Environment

Fire spreads by feeding on fuel, which is to say, the biomass of living landscapes. Change that fuel array and you change fire. We can't rearrange mountains, redirect winds, or ban droughts; we can rearrange trees, shrubs, windfall, litter, and grass. This of course is the premise of agriculture: remake the environment into forms more suited to people's desires. And it is the foundation of the built landscape, every part of which is (in theory) under the control of people.

This is another legacy strategy: it is how people in fire-prone settings have survived across history. In its simplest form, changing the fuel array can mean preemptive burning that removes flammable dead combustibles and promotes less flammable living ones. But it can mean cultivating fields and pastures and reorganizing the living landscapes into gardens, arable acres, and paddocks. It can refer to cities built out of noncombustible materials, fashioned from synthetics or organized to suit human economies rather than those of nature and then arranged in ways to discourage fire spread. The tighter the control over what exists, the stronger the control over fire.

The oldest, most intensively inhabited lands of Europe are those along its southern rim, within a Mediterranean climate, which is practically a dictionary definition of a fire-prone setting. These lands have survived by close cultivation and herding, including the use of fire to remove stubble or fallow or to freshen browse: little fires, frequent fires; field fires, horticultural fires; fires to remove slash and cuttings, fires to prune berries, fires to clear the ground prior to harvesting olives and nuts. The fires are easily controlled because they are small, burn amid tightly cultivated fuels, and have people at hand to swat out any errant embers. Wildfires occur, but only when the social order breaks down from war, riot, famine, or plague; when the landscape is no longer so meticulously maintained; when the garden goes to seed and the field to waste. Fire follows the havoc. In such settings fire is an index of social order.

The strategy undergirds European thinking about fire. It's especially pertinent in temperate Europe, since fire has little natural presence and only exists because people put it in (or take it out).

The social order determines the character of the countryside, and the countryside's combustibles determine what kind of fire can burn. All this helps explain Europe's conception of fire as a tool and a measure of human presence. It accounts for why European agronomists long disregarded fire ecology in favor of social models and why America's first professional forester, Bernhard Fernow, a Prussian émigré, dismissed the entire American fire scene as one of "bad habits and loose morals." It explains why Mediterranean Europe—a fire-avid place depopulated by a modern economy—hosts almost all the continent's wildfires.

It also helps explain arguments for active land management as a fundamental axiom of fire protection. Thinned forests, culled shrublands, dormant grasses grazed or burned, all can affect how fire behaves and with what consequences. But interventions can also worsen the fire scene. Logging leaves slash, as volatile a fuel as any on the planet. Overgrazing can destroy perennial grasses and help convert a landscape to fire-hungry weeds like cheatgrass, or it can allow woody plants like eastern red cedar to invade or delete the needful effects of fire. As with all tools and practices, the outcome depends.

It depends first on whether the interventions act on factors that fire understands. Fire is sensitive to the small, the quickly kindling, and the connective tissues of combustibles that link shrubs with bunch grasses, surfaces to crowns, one canopy to another. The big and the wet, no matter how massive, won't carry flame; not all biomass is available as fuel. Logging, for example, takes the big stuff and leaves the little; fire burns the little and leaves the big. What remains after even the most savage wildfire are the trunks that logging would have removed.

Yet just as there is a place for prescribed burning, so one can imagine prescribed cutting and crushing, prescribed grazing and browsing, and prescribed planting and harvesting. Whether those interventions are helpful depends on their scale and purpose, on whether they enhance ecological integrity or simply reduce complex ecosystems to blocks of hydrocarbons and bulk commodities. Mediterranean Europe's fire problems could be significantly reduced by reinstating an updated version of the cultivation and herding that had for centuries shaped its landscapes. It could treat its built landscape (one made of biotic materials) as though it were a heritage building that retains its old form and facade but has its interior outfitted with modern furnishings. The wildlands of western North America could benefit from modern variants of the aboriginal practices with which humans had managed to live with fire for millennia. What won't work is relying on bigger machines to counter the feral fires that will inevitably arrive if the land is untreated and all controlled fires are denied.

Exclude Fire

The last option is to remove fire as a presence. Put simply, this means preventing fires from starting and suppressing those that do ignite before they can spread. Stopping fires is the flip side to starting them; without the capacity to influence how fires spread, kindling flames can be indistinguishable from vandalism. Traditionally, the only way to halt propagating fires was to swat out the flames, break the continuity of fuels, and set counter fires. Even stopping fires required setting them; burning out could resemble other varieties of controlled fire.

The pyric transition washed over these fires as it did every other fire practice. It replaced open flames with internal combustion used to swat out flames with water and retardants and to remove fuel with mechanical plows, blades, and saws. It replaced the counter fire of back burning with the counterforce of industrial combustion. In cities, all this worked to the good. In the countryside, particularly in wildlands, it created an illusion of protection, perhaps of political theater, that suggested an appearance of control even as it encouraged firescapes to spiral out of all restraint.

One on one, industrial combustion could never match the pervasive fire power of free-burning flame. It was instrumental in changing fuels and climate broadly. And it was effective at small fires—at getting water, dirt, and personnel to fires that remained tiny. It was ineffective against the droughts and the high winds that powered big burns. Suppression helped nature's fire economy polarize into a proletariat of many small fires and a few plutocratic blowups (the 1 percenters) that racked up most of the burned area, accounted for the greater percentage of damages, and rang up the biggest bills.

The problem is not suppression as a concept, but the appeal to suppression as a sole-source strategy for responding to fire outside built environments. It can help—for a while—to keep bad fires out of a landscape; it cannot keep good fire in. While it promises better control, it can succeed only where there is no natural foundation for fire, where fire's presence is wholly within the hands of humans. It can only assist fire control if people also have control over the firescape. Otherwise, it can disrupt without installing a workable replacement. It is a strategy for putting down an ecological riot, not for governing.

The consequences of fire exclusion may be disguised for years, even decades, until a new generation assumes that the nature it knew in its childhood is the nature that was meant to be and the former regime is no longer visible. The effects show soonest where fire regimes are characterized by frequent surface fire through grasses and low shrubs—not just prairie, but also oak-hickory woodlands and savannas, and ponderosa and longleaf pine forests that overgrow with reproduction, which can carry flame from the surface to the crown, or flood with less fire-tolerant species and thicken with fire-sensitive trees that alter the microclimate. The effects manifest more slowly where burns are infrequent but accelerate to crown fires, such as among lodgepole and jack pine and among black spruce as they lose patchiness in favor of even-aged swards that can support long fetches and fields of unbroken fire.

Practice Rather Than Perfection

Like fire, which rarely does just one thing, fire strategies rarely stay within boxes and file folders, and as fires become more varied, so must responses. They merge and muddle the clarity that modern management aspires to. More and more, fire's management resembles a mash-up. With so much of the environment changing so rapidly, regimes are mingled, responses mixed, firescapes scrambled, and knowledge syncretic. While fire's power to integrate makes it difficult to model, it also means that there are many points of intervention possible, and the future will require cocktails of treatments adjusted to particular places.

What were once stand-alone, singular practices are now plural. Prescribed fire includes set-piece burns on agricultural and Florida models, but also large burnouts done as part of suppression. There are cultural burns, ecological burns, agricultural burns, hazard reduction burns, protective burns, box-and-burns. So, too, suppression now includes towns, countryside, and wildlands, all with their distinctive equipment, tactics, purposes, and fire cultures. Each has variable definitions for what *control* means. It can mean extinguishment of every ember, or control over the perimeter, or control by confining a burn to a particular stretch of land. Firescapes can include nature reserves, recreational sites, legal wilderness, plantations, dispersed housing, and trailer parks. Scientific understanding has to expand to include disciplines not traditionally considered the purview of landscape fire; to incorporate traditional ecological knowledge, from the Wurundjeri in Australia to the Karuks in California; and to accept the lore of working fire cultures like the pastoral burning in the Flint Hills of Kansas, the open-range burning in Florida, and the habitat burning in the Red Hills of the southeastern coastal plain.

All disrupt the clarity and certainty promised by the previous ideal, in which science informed and management applied. Instead, field science is learning by doing. There are fire managers on the ground who are challenging computer simulations based on putatively first principles. The basic questions are not even amenable to science but reside in social, cultural, and political spheres, which means incorporating art, philosophy, literature, and law. Fire synthesizes its surroundings: so, too, fire's management is having to replace the reductionism that is the marvel of modern science with the integrating power that is inherent in fire. We can't find separate surrogates for everything fire does. We don't need to, because fire will do that for us. We just have to get a reasonable amount of good fire on the ground. If we let it, fire will sieve and sort and synthesize for us.[12]

Likewise, we don't need a dedicated fire program to address all the ills of the Pyrocene. Fire is interactive; it can bring to human economies the broad-spectrum spark it brings to nature's. Fire crises can catalyze reforms that were needed anyway. Coal-fired power plants have worsened air quality, apart from greenhouse gases; hilltop mining for coal has trashed landscapes. A creaky power grid has needed rehabilitation, apart from its ruinous sparks during high winds. Urban sprawl has been a social and land use problem, apart from putting communities at risk from fire. Invasive species, extinguished species, fragmented habitat—all were urgent issues before climate change. Nature reserves needed more good fire even under the climate of the mid-twentieth century. We can direct the alarm aroused by unchecked burning to rally the will to take on tasks we have too long deferred. Fire is plural, fire is systemic. It has no single solution, and never will, but it can give urgency and focus to how humans choose to live in the coming epoch.

Fire isn't ecological pixie dust that sprinkled willy-nilly over the landscape will magically restore or make everything right. But it can help bring what exists into a working whole. That doesn't ensure a world we like or need. If we want that world to be habitable for humans, we will have to put our hand and mind on the firestick and select for fires that serve our visions and needs better. That requires social investments, political capital, and a reconstructed relationship to fire.

We will have to accept that fire is not a fringe phenomenon but an informing principle of terrestrial life and human culture. This does not mean that everything will or must burn. It does mean that most everything will be affected, if indirectly, by humanity's fire habits. Not all of Earth in the Pleistocene was buried under ice sheets, but not much of the planet was unaffected by the ice. What ice was to the Pleistocene, fire is to the Pyrocene. The Earth has long been a fire planet. It is now advancing into a deepening fire age.

Epilogue

Sixth Sun

The Cerro de la Estrella rises in the center of the Valley of Mexico. In pre-Columbian times, it was an island, where the waters of Lake Texcoco mingled with Lake Xochimilco. It was here, every 52 years, that the Aztecs celebrated the ceremony of the New Fire. Then, when the two calendars, the 260-day and the 365-day, coincided, when the Pleiades stood overhead, when the cosmos was poised to crash into darkness or to rekindle into a reborn light, the New Fire redeemed the world and birthed a new sun.[1]

The ceremony was elaborate, its siting on the cerro majestic. In all the countryside around, across the wide-mirrored lake, in every hearth and village, in every temple, in every torch and campsite, the fires were extinguished until all human light had vanished from the evening darkness. Only the illumination of the stars remained. The world—the known world of the sun—shuddered in uncertainty. The dark and the demons crept closer. Only a renewed fire, kindled in the ancient way, the way humans first learned to make it, could spark the sun's return.

On an altar at the cerro's summit, four priests waited, one for each of the elements, for each of the four previous worlds, for each of the four 13-year counts whose beats summed up to a New Fire. A fifth priest ripped out the beating heart of a prisoner, the mandatory human sacrifice. A New Fire emerged from the sacred implements and was placed in the exposed chest to signify new life; then each of the four priests ignited a great torch from the common New Fire and, surrounded by guards, marched down the slopes to boats waiting to take them to each of the cardinal points. Ashore, the priests kindled the fuels of a subsidiary New Fire, each overseen by a priestess whose task it was to keep the fire burning for another 52 years. To fail was fatal. From this fire, all the fire of hearths, furnaces, and temples; all the fires used in hunting and fishing; all the fires of life sacred and profane were rekindled. The stars would wheel in their places. The sun would rise. Once more, the world would be saved.

Five times the world had ended. Five times a new sun had eventually risen. The success of the ceremony atop Cerro de la Estrella assured the continuance of the existing world of the Fifth Sun. The last ceremony occurred in 1507. Before the next cosmic convergence of time and space could occur, newcomers arrived from the east, from the realm where the sun rose daily, and allied with Indigenous armies, they destroyed the Aztec empire. The New Fire ceremony ceased.

The destruction that led to a Sixth Sun, however, had to await another visitation from the east. This one brought fire that kindled out of the heart of the living world, a New Fire that burned fossil fuel and that, like its symbolic predecessors, required that all other fires be extinguished while it disseminated to all the corners and crevices of the Earth.

·   ·   ·

The Pyrocene began when a fire-wielding creature met a fire-receptive period in the Earth's history and their interaction made anthropogenic fire an informing presence. There is a case for arguing that the epoch spans the whole of the Holocene, that while its pace quickened with the advent of the wholesale burning of fossil fuels, the narrative of humanity's career as keeper of the planetary flame is continuous and unbroken. There is also a case for a briefer Pyrocene in which the pivot to fossil biomass marks a phase change in kind, not just quantity. By that thinking, it is unfair to lump all the more or less judicious uses of fire in living landscapes with the global rupture sparked by burning lithic landscapes, unfair to condemn all of humanity when only a small fraction was responsible for unleashing the combustion cascade that has washed over the planet.

When a narrative should start depends on when it ends. A *long* Pyrocene has the advantage of showing how today's binge-burning happened, of reminding us that we had a fire crisis before climate change threw accelerant onto the flames, that fire will not go away when the climate stabilizes or reverts to earlier forms, that fire is what we as a species do. A *short* Pyrocene avoids arguments about origins, distractions over dominant drivers behind bad burns, and misdirections about the proportional causation of megafires and megapalls, and it lasers in on the disruption propagated by industrial combustion. Either version identifies Earth's keystone species for fire as the spark behind the burning. People and climate interact amid complex feedbacks that can parley two into four, and four into sixteen.

I favor a long Pyrocene because it testifies to the in-our-genome bond between humans and fire and because it provides a deep history for what promises to be a deep future. A warming Earth did not require people to start the process, but it seems likely it will require people in order to continue.

Of all the paradoxes proposed by the Pyrocene, the strangest may be that our fire practices may have unwittingly forestalled the return of the ice. The Little Ice Age might have continued; the next, promised glacial might be pushed to the margins as humans fiddle with climatic rheostats. Since it's easier for people to live with fire than with ice, we may have bought ourselves some time—some room for maneuver. Our fire practices, however inadvertently, may have spared us an impossible future of ice, though we may perish equally from fire if we don't control our burning, which is to say, ourselves.

We need to keep our fossil biomass in the ground. We need to store it, not only to cool humanity's feverish presence today, but to have it as a stockpile to ward off the ice's return in the future. It's the climate equivalent of a strategic petroleum reserve. It's our hedge against the coming cold. We may need to burn a lot of it, and when the ice approaches, we'll be glad for every particle of carbon-dense combustibles we saved when the world was warm. We will need to exercise our fire power of all kinds in perpetuity. We will remain the keeper of the flame until we end.

. . .

But what flame are we keeping? How we imagine fire will inform the relationship we have with it.

We can define fire as a chemical reaction shaped by its physical surroundings and look to physical measures to exploit and contain it. We can house it in physical settings and machines. We can try to overcome its natural expressions with physical countermeasures. We can contain it within incombustible stone, water, and dirt barriers. We can dump water and retardants on it. Escaped, it resembles a tsunami or hurricane.

Or we can define it as fundamentally biological; not itself living but, like a virus, a process that depends on a biotic context and biochemical reaction to propagate. We can direct it by domesticating it, taming it, creating an ecological setting in which the fire will do what we want and not do what we don't. We engineer it through ecological means. Its escapes result from broken biotas; outbreaks resemble a contagion, and megafire an emergent plague.

Or we can envision the whole of our relationship as culturally grounded. After all, we decide what fires are good and bad, which flames belong in torches and which in fields, what ideas and institutions are appropriate to manage it. The whole notion of a relationship is ours, not fire's. Breakdowns result from human activity; it is our behavior, not fire's, that is troubled.

Which correctly characterizes fire? Ever the shape-shifter, fire is all of the above, and more, limited only by the capacity of Earth and humans to invent contexts. Each conception has its place. For built landscapes, machines, and fire that is blowing in the wind, a physical model is appropriate. For living landscapes, either cultivated or wild, biological fire is suitable, and treating the scene through a physical model a problem. For a fire scene like the Pyrocene, fashioned by our practices and attitudes, only the cultural model can address the core drivers. Only it can create and choose among possible narratives and the behavior they invite.

· · ·

Yes, I do know my origin!
Unquenchable like the flame
I consume myself and glow.
Into light turns all I touch,
Into coal all that I leave:
Flame is surely what I am!

FRIEDRICH NIETZSCHE, *Ecce Homo*

Praised be You, my Lord, through Brother Fire, through whom thou givest
us light in the darkness: and he is bright and pleasant and very mighty
and strong.

ST. FRANCIS OF ASSISI, *Canticle of the Creatures*

In the grand saga of the Pyrocene, until recent centuries, it was enough that humanity pursued its quest for fire, that it endlessly sought more stuff to burn in more places. Two complementary narratives walked together along parallel paths. The Promethean (or Nietzschean) narrative spoke to fire as power. It cast anthropogenic fire as something wrenched from its natural setting, perhaps by violence, and redirected to human purposes. The primeval (or Franciscan) narrative conceived fire as a companion on our journey, as something for which humans were stewards, managing for their well-being as well as that of the rest of creation. In both narratives fire was what humans did that no other creature could. In one, humans hoarded; in the other, they shared. The two conceptions could range widely, but their paths remained within a common realm, the dominion of living landscapes.

With fossil fuel combustion the paths parted. With increasing intensity and momentum, humanity left the Franciscan conception for the Nietzschean, and the living landscape for the lithic. There was less primeval fire and more Promethean until the Promethean, now unchained, commanded the heights. Humanity's fire power spiked at exponential rates, but the living world in which we live suffered and threatened to become increasingly uninhabitable. Now we are poised to exchange a full-blown ice age for a runaway fire age.

We need a lot less Promethean fire and a lot more primeval. We need to recover cultural burning at a species level. We need to remember that fire is not simply a tool, a presence, or a process for us to manipulate, but a relationship; that we cannot exist without fire, but fire can exist without us; that our unique fire power brings with it a unique responsibility.

We have used our fire power to reset the world and birth a Sixth Sun. Now we have to learn how to distribute that flame for the greater good of the Earth because its future is also our own.

Author's Note

This extended essay distills and reformulates a lifetime of engagement with fire. But while fire is infinitely malleable—a shape-shifter, nature as postmodernist— I am not, nor are my words. I have found only so many ways to express what I understand, so this brief book repeats or paraphrases sentences, paragraphs, and passages from previous works. In particular, *The Pyrocene* builds on *Fire: A Brief History,* 2nd ed. (University of Washington Press, 2019), supplemented by *The Last Lost World* (Viking, 2012) and *The Ice* (University of Iowa Press, 1986), though which I triangulate my understanding of the ice ages. Probably 80 percent of what this text addresses can be found in some form in my previous publications. (This makes the fourth book in which I have addressed the four strategies of fire management, though each version has come with its own thematic rebars and idiom.) What I have added in this book is a new organizing concept that I trust grants those thoughts and words an enriched context and fresh meaning.

I coined the term *Pyrocene* as a catch phrase in a 2015 essay titled "The Fire Age" published in *Aeon* (May 5, 2015, https://aeon.co/essays/how-humans-made-fire-and-fire-made-us-human). Since then I've started using it regularly and in 2019 began to propose it as an informing principle (in a literary sense) by which to understand the world our pact with fire has made. I have appealed to it particularly where the living and the lithic landscapes have crossed in ways that exploded into flame or where they interacted to reconstruct land-scapes. Think of the power lines that have kindled so many of California's bad fires; Alaska, where a petrostate supports people living in a fire-prone boreal landscape; or Fort McMurray, Alberta, where a community created to mine oil sands is overrun by fires, probably leveraged by climate change, rushing out of the surrounding woods. Once again, *Aeon* (November 19, 2019, https://aeon .co/essays/the-planet-is-burning-around-us-is-it-time-to-declare-the-pyrocene) provided a venue for a Pyrocene piece, publishing just as Australia's record 2019–20 fire season began hitting stride.

There are many names for the Age of Humans. They all have their value, each emphasizing particular causes and consequences. In time, one phrase will surely overtop the others, like resprouting stems that yield to one as dominant. From a geologic perspective, I have long regarded all of the Holocene as an Anthropocene. From a fire perspective, I now regard the Anthropocene as a Pyrocene.

I would like to thank TED Talks for forcing me to condense my understanding of fire history into 14 minutes, Brigid Hains at *Aeon* for granting me a medium in which to convert that performance into words and then return with an alternative narrative, and various editors at *Slate, History News Network, Natural History, Fire,* and the *Guardian* and endless journalists seeking commentary on fires in the news, all of whom forced me to push myself into sharper observations, unexpected insights, and a punchier prose.

And as always thanks to Sonja, this time for insisting I pursue *The Pyrocene,* which she saw as a culmination of my long personal quest for fire.

Notes

Prologue

1. V. Alaric Sample, R. Patrick Bixler, and Char Miller, eds., *Forest Conservation in the Anthropocene: Science, Policy, and Practice* (Boulder: University Press of Colorado, 2016); V. Alaric Sample and R. Patrick Bixler, eds., "Forest Conservation and Management in the Anthropocene: Conference Proceedings," Proceedings, RMRS-P-71, US Department of Agriculture, Forest Service, 2014, https://www.fs.usda.gov/treesearch/pubs/46127.

2. For a brief explanation of the fire paradox, see Mark Finney's video at https://wildfiretoday.com/2018/03/28/the-fire-paradox. On the lessening of burned area globally, see N. Andela et al., "A Human-Driven Decline in Global Burned Area," *Science* 356 (2017): 1356–1362.

Chapter 1. Fire Planet

1. Clinton B. Phillips and Jerry Reinecker, "The Fire Siege of 1987: Lightning Fires Devastate the Forests of California," California Department of Forestry and Fire Protection (Sacramento, 1988); California Department of Forestry and Fire Protection, "2008 Wildfire Activity Statistics," www.fire.ca.gov/media/10885/2008_wildfireactivitystatistics_complete_revised.pdf.

2. This paragraph quotes or paraphrases passages from Stephen J. Pyne, *Fire: A Brief History*, 2nd ed. (Seattle: University of Washington, 2019), 8. See Richard Blaustein, "The Great Oxidation Event," *Bioscience* 66 (March 2016): 189–95, and Andrew C. Scott, *Burning Planet* (New York: Oxford University Press, 2018).

3. James Lovelock, *The Ages of Gaia* (New York: Bantam Books, 1988), 29.

4. This paragraph quotes or paraphrases passages from Pyne, *Fire: A Brief History*, 2nd ed., 9.

5. See Juli G. Pausas and William J. Bond, "On the Three Major Recycling Pathways in Terrestrial Ecosystems," *Trends in Ecology & Evolution* 35(9) (September 1, 2020): 767–775.

6. This paragraph quotes or paraphrases passages from Pyne, *Fire: A Brief History*, 2nd ed., 10–11.

7. The fire regime concept is fundamental to fire ecology. An interesting variant is the concept of the "pyrome," which is to fire as biome is to ecology. The term has not yet caught on, though it might (and should). See Sally Archibald et al., "Defining Pyromes and Global Syndromes of Fire Regimes," *Proceedings of the National Academy of Sciences* 110(16) (April 16, 2013): 6442–6447, www.pnas.org/cgi/doi/10.1073/pnas.1211466110.

8. Ashley Strickland, "A Dinosaur's Last Meal: A 110 Million-Year-Old Dinosaur's Stomach Contents Are Revealed," CNN (June 2, 2020), www.cnn .com/2020/06/02/world/nodosaur-fossil-stomach-contents-scn-trnd/index .html.

9. This paragraph quotes or paraphrases passages from Pyne, *Fire: A Brief History*, 2nd ed., 15.

10. As illustrations, see Leda N. Kobziar et al., "Pyroaerobiology: The Aerosolization and Transport of Viable Microbial Life by Wildland Fire," *Ecosphere* 9(11) (November 2018): article e02507; Elizabeth Thompson, "Wildfire Smoke Boosts Photosynthetic Efficiency," *Eos* 101 (February 12, 2020), https://doi .org/10.1029/2020EO139985; Manoj G. Kulkarni and Johannes Van Staden, "Germination Activity of Smoke Residue in Soils Following a Fire," *South African Journal of Botany* 77 (2011): 718–724; Matthew W. Jones et al., "Fires Prime Terrestrial Organic Carbon for Riverine Export to the Global Oceans," *Nature Communications* 11 (2020): article 2791, https://doi.org/10.1038/s41467-020-16576-z.

11. For a very accessible introduction to these terms, see Ronald L. Myers, *Living with Fire: Sustaining Ecosystems and Livelihoods through Integrated Fire Management* (The Nature Conservancy, 2006), 3–6.

12. Jeff Hardesty, Ron Myers, and Wendy Fulks, "Fire, Ecosystems, and People: A Preliminary Assessment of Fire as a Global Conservation Issue," *The George Wright Forum* 22(4) (2005): 78–87.

13. I rely mostly on Scott, *Burning Planet,* and chapters 3 and 4 in Andrew Scott et al., *Fire on Earth: An Introduction* (Chichester: Wiley-Blackwell, 2013).

14. This paragraph quotes or paraphrases passages from Pyne, *Fire: A Brief History,* 2nd ed., 12.

15. The original conception is well summarized in Walter Alvarez, *T. Rex and the Crater of Doom* (Princeton: Princeton University Press, 1997). Recent research presents a more modulated relationship between impact and residual charcoal; see, for example, "Chicxulub Crater Reveals the Terrible End of the Dinosaurs," *Inverse* (December 15, 2019), www.inverse.com/article/61695-chicxulub-crater-reveals-end-of-dinosaurs, and "Earth's Most Destructive Day Ever: Chicxulub Crater Evidence Study Tells a New Story," *Inverse* (September 9, 2019), www.inverse.com/article/59122-chicxulub-crater-study-reveals-wildfires-tsunamis.

16. See Dag Olav Hessen, *The Many Lives of Carbon* (London: Reaktion Books, 2017), 178–179, for PETM, and Scott, *Fire on Earth,* 75–76 and 88, for fire.

17. This paragraph quotes or paraphrases passages from Pyne, *Fire: A Brief History,* 2nd ed., 12–13.

18. For a brief introduction to the intellectual history of fire in the West, see Stephen Pyne, "Fire in the Mind: Changing Understandings of Fire in Western Civilization," *Philosophical Transactions of the Royal Society B* 371 (2016): 20150166, https://doi.org/10.1098/rstb.2015.0166.

19. William Crookes, ed., *Course of Six Lectures on the Chemical History of a Candle* (London: Griffin, Bohn, and Co., 1861).

20. William James, *The Varieties of Religious Experience* (New York: Longmans, Green, and Co., 1917), 74.

Chapter 2. The Pleistocene

1. A lively recapitulation is available in John Imbrie and Katherine Palmer Imbrie, *Ice Ages: Solving the Mystery* (Cambridge: Harvard University Press, 1986), 19–31. But see also Edward Lurie, *Louis Agassiz: A Life in Science* (Chicago: University of Chicago Press, 1960).

2. William F. Ruddiman, *Plows, Plagues and Petroleum* (Princeton: Princeton University Press, 2005), 41.

3. Ruddiman, *Plows,* 121. There are many dates proposed for the Little Ice Age, ranging from 1350 to 1900. A consensus seems to exist around 1850 for an

end date, though some researchers argue for a longer tail. I agree with most researchers on 1550 as a beginning date, though the evidence is not conclusive. A cooling set in after the Medieval Warm Period, but when that announces a Little Ice Age depends on criteria and themes.

4. Ruddiman, *Plows*, 84–86.

5. Elaine Anderson, "Who's Who in the Pleistocene: A Mammalian Bestiary," in Paul S. Martin and Richard G. Klein, eds., *Quaternary Extinctions: A Prehistoric Revolution* (Tucson: University of Arizona Press, 1984), 40–89, and Paul S. Martin, "Prehistoric Overkill: The Global Model," 354–403, in the same book. The overkill model has been extensively modified over the years, but note especially the graphs on page 395. Note, too, that at the time Martin and Klein published, the Pleistocene encompassed 1.9 million years. Subsequently the epoch has expanded to 2.6 million years, which also enlarges the scope of its extinctions.

6. This paragraph and the four that follow quote from or paraphrase Lydia V. Pyne and Stephen J. Pyne, *The Last Lost World: Ice Ages, Human Origins, and the Invention of the Pleistocene* (New York: Viking, 2013), 30–33.

7. A useful survey of the history of Pleistocene chronological definitions is in J. J. Low and M. J. C. Walker, *Reconstructing Quaternary Environments* (New York: Longman, 1984), 3–8.

8. To follow the debate, see Imbrie and Imbrie, *Ice Ages*, 123–173; Paul E. Damon, Glen A. Izett, and Charles W. Naeser, conveners, "Pliocene and Pleistocene Geochronology," Penrose Conference Report, *Geology* 4 (October 1976): 591–593; and Amanda Leigh Mascarelli, "Quaternary Geologists Win Timescale Vote," *Nature* 459 (June 4, 2009): 624.

Chapter 3. Fire Creature: Living Landscapes

1. Quoted in Josephine Flood, *Archaeology of the Dreamtime: The Story of Prehistoric Australia and Its People* (Sydney: Angus and Robertson, 1999), 227. See also T. L. Mitchell, *Journal of an Expedition into the Interior of Tropical Australia* (London, 1848), 306, available at www.gutenberg.org/files/9943/9943-h/9943-h.htm; T. L. Mitchell, *Three Expeditions in the Interior of Eastern Australia* (London, 1839), 196, available at www.gutenberg.org/files/12928/12928-h/12928-h.htm (Vol. 1) and http://gutenberg.net.au/ebooks/e00036.html (Vol. 2).

2. Mitchell, *Journal of an Expedition*, 412.

3. Mitchell, *Journal of an Expedition*, 413.

4. See Richard Wrangham, *Catching Fire: How Cooking Made Us Human* (New York: Basic Books, 2009).

5. See Konrad Spindler, *The Man in the Ice* (London: Phoenix Books, 1993); Samir S. Patel, "Illegally Enslaved and Then Marooned on Remote Tromelin Island for Fifteen Years, with Only Archaeology to Tell Their Story," *Archaeology* (September/October 2014), www.archaeology.org/issues/145-1409/features /2361-tromelin-island-castaways#art. See also Mich Escultura, "The Inspiring Story of the Castaways of Tromelin Island," *Elite Readers* (October 11, 2016), www.elitereaders.com/castaways-tromelin-island, and "Lèse humanité," *Economist* (December 16, 2015), www.economist.com/christmas-specials/2015/12/16 /lese-humanite.

6. For examples of fire as micromanagement, see Kat Anderson, *Tending the Wild: Native American Knowledge and the Management of California's Natural Resources* (Berkeley: University of California Press, 2013).

7. Rhys Jones, "Fire-Stick Farming," *Australian Natural History* 16 (1969): 224–228; Bill Gammage, "Australia under Aboriginal Management," Fifteenth Barry Andrews Memorial Lecture, University College, Canberra, 2002, and *The Biggest Estate on Earth: How Aborigines Made Australia* (Sydney: Alley & Unwin, 2012).

8. Rhys Jones, "The Neolithic, Palaeolithic, and the Hunting Gardeners: Man and Land in the Antipodes," in R.P. Suggate and M.M. Cresswell, eds., *Quaternary Studies* (Wellington, 1975), 26.

9. For an extraordinarily intensive investigation, see William Baleé, *Footprints of the Forest: Ka'apor Ethnobotany—the Historical Ecology of Plant Utilization by an Amazonian People* (New York: Columbia University Press, 1994), especially 136–138 and 220–222.

10. The best summary for Europe is Stephen J. Pyne, *Vestal Fire: An Environmental History, Told through Fire, of Europe and Europe's Encounter with the World* (Seattle: University of Washington Press, 1997), but that derives from many other sources, among them Axel Steensberg, *Fire Clearance Husbandry: Traditional Techniques Throughout the World* (Herning: Poul Kristensen, 1993); François Sigaut, *L'Agriculture et le feu: Role et place du feu dans les techniques de préparation du champ de l'ancienne agriculture européenne* (Paris: Mouton & Co., 1975); and for Finland, the special issue of *Suomen Antropologi* 4 (1987). The great compendium on swidden and fire for tropical lands (broadly drawn) is Harley H. Bartlett, "Fire in Relation to Primitive Agriculture and Grazing in the

Tropics: Annotated Bibliography," Supplement to Background Paper No. 34, "Man's Role in Changing the Face of the Earth," a summary of which is available as "Fire, Primitive Agriculture, and Grazing in the Tropics," in William L. Thomas Jr., *Man's Role in Changing the Face of the Earth,* vol. 2 (Chicago: University of Chicago Press, 1956), 692–720.

11. On fire-forage grazing in Europe, see Pyne, *Vestal Fire,* which includes examples for each of the five fire provinces of Europe proper. For a nice survey of transhumance, see Elwin Davies, "Patterns of Transhumance in Europe," *Geography* 26 (1941): 116–127.

12. Quoted in Cyril Stanley Smith and Martha Teach Gnudi, trans. and eds., *The Pirotechnia of Vannoccio Biringuccio* (Cambridge: MIT Press, 1966; New York: Dover, 1990, reprint), xxvii.

13. On the Mexico discoveries, see Ciprian F. Ardelean et al., "Evidence of Human Occupation in Mexico around the Last Glacial Maximum," *Nature* (July 22, 2020), https://doi.org/10.1038/s41586-020-2509-0.

14. Statistics on grasslands are notoriously sensitive to definitions, not least what constitutes a grassland (as distinct from a woodland or savanna). I found the following useful, though dated, as compendiums: Robin P. White, Siobhan Murray, and Mark Rohweder, *Grassland Ecosystem* (Washington, DC: World Resources Institute, 2000), and Eleonora Panunzi, "Are Grasslands Under Threat? Brief Analysis of FAO Statistical Data on Pasture and Fodder Crops," www.fao.org/uploads/media/grass_stats_1.pdf.

15. Sourced from World Bank data, https://data.worldbank.org/indicator/AG.LND.AGRI.ZS?end=2016&start=1961.

16. The shift in expected rhythms is wonderfully explained in William F. Ruddiman, *Plows, Plagues, and Petroleum: How Humans Took Control of Climate* (Princeton: Princeton University Press, 2005). I'm elaborating on his argument by including fire-specific effects.

17. The classic study remains Jean M. Grove, *The Little Ice Age* (London: Methuen, 1988). A popular presentation of the long summer concept is a book with that title: Brian Fagan, *The Long Summer: How Climate Changed Civilization* (New York: Basic Books, 2004).

18. On human demographics and the onset of the Little Ice Age, see Robert A. Dull et al., "The Columbian Encounter and the Little Ice Age: Abrupt Land Use Change, Fire, and Greenhouse Forcing," *Annals of the Association of American Geographers* 100(4) (2010): 755–771, https://doi.org/10.1080/00045608

.2010.502432. For an update, see Alexander Koch et al., "European Colonisation of the Americas Killed 10% of World Population and Caused Global Cooling," *The Conversation* (January 31, 2019), https://theconversation.com/european-colonisation-of-the-americas-killed-10-of-world-population-and-caused-global-cooling-110549.

Chapter 4. Fire Creature: Lithic Landscapes

1. Alexander Napier, ed., *The Life of Samuel Johnson, LL.D. together with the Journal of a Tour to the Hebrides by James Boswell, Esq.,* vol. 3 (London: George Bell and Sons, 1884), 42.

2. Napier, *Life of Samuel Johnson,* 62.

3. The Odum quote is from Howard T. Odum, *Environment, Power, and Society* (New York: Wiley Interscience, 1970), 116.

4. N. Andela et al., "A Human-Driven Decline in Global Burned Area," *Science* 356 (June 30, 2017): 1356–1362.

5. The wildland-urban interface is a pathology of developed countries; studies exist for America, Australia, France, and Canada, and there are separate inquiries in response to tragedy fires for others. For a detailed cartographic survey as a sampler, see Sebastíin Martinuzzi et al., "The 2010 Wildland-Urban Interface of the Conterminous United States," Research Map NRS-8 (Newtown Square, PA: US Department of Agriculture, Forest Service, 2015), https://doi.org/10.2737/NRS-RMAP-8.

6. Two canonical works on deforestation and European thinking are Richard Grove, *Green Imperialism: Colonial Expansion, Tropical Island Edens, and the Origins of Environmentalism, 1600–1860* (Cambridge: Cambridge University Press, 1995), and Michael Williams, *Deforesting the Earth: From Prehistory to Global Crisis* (Chicago: University of Chicago Press, 2002).

7. A.A. Brown and A.D. Folweiler, *Fire in the Forests of the United States* (St. Louis: John S. Swift Co., 1953), 3.

8. S.B. Show and E.I. Kotok, "The Role of Fire in the California Pine Forests," Department Bulletin No. 1294, US Department of Agriculture (Government Printing Office, 1924), 47.

9. Examples are legion, but I find the story in Ethiopia particularly revealing, not least because it seems to recycle the California story a century later. See Maria Johansson, Anders Granström, and Anders Malmer, "Traditional

Fire Management in the Ethiopian Highlands: What Would Happen If It Ends?" *Forest Facts* 9 (2013), results from the Swedish University of Agricultural Sciences.

Chapter 5. The Pyrocene

1. See MNP LLP, "A Review of the 2016 Horse River Wildfire: Alberta Agriculture and Forestry Preparedness and Response," prepared for the Forestry Division, Alberta Agriculture and Forestry, Edmonton (June 2017), www.alberta.ca/assets/documents/Wildfire-MNP-Report.pdf.

2. M. Turco, S. Jerez, S. Augusto, et al., "Climate Drivers of the 2017 Devastating Fires in Portugal," *Scientific Reports* 9 (2019): article 13886, https://doi.org/10.1038/s41598-019-50281-2.

3. California Department of Forestry and Fire Protection, Camp fire summary, www.fire.ca.gov/incidents/2018/11/8/camp-fire. See also Alejandra Reyes-Velarde, "California's Camp Fire Was the Costliest Global Disaster Last Year, Insurance Report Shows," *Los Angeles Times* (January 11, 2019), www.latimes.com/local/lanow/la-me-ln-camp-fire-insured-losses-20190111-story.html.

4. See Dave Owens and Mary O'Kane, *Final Report of the NSW Bushfire Inquiry* (Sydney: NSW Government, July 31, 2020). For a well-illustrated account of the largest of the fires, Gospers Mountain, see Harriet Alexander and Nick Moir, "'The Monster': A Short History of Australia's Biggest Forest Fire," *Sydney Morning Herald* (December 20, 2019), www.smh.com.au/national/nsw/the-monster-a-short-history-of-australia-s-biggest-forest-fire-20191218-p53l4y.html.

5. Elizabeth B. Wiggins et al., "Smoke Radiocarbon Measurements from Indonesian Fires Provide Evidence for Burning of Millennia-Aged Peat," *Proceedings of the National Academy of Sciences* 115(49) (December 4, 2018): 12419–12424, https://doi.org/10.1073/pnas.1806003115.

6. See Bruce Finley, "Wildfire Haze, Record Heat and Pollution Combine to Make Denver Air Quality Dangerous for All," *Denver Post* (August 25, 2020), www.denverpost.com/2020/08/25/colorado-wildfire-smoke-pollution-ozone.

7. Jennifer K. Balch et al., "Human-Started Wildfires Expand the Fire Niche across the United States," *Proceedings of the National Academy of Sciences* 114(11) (February 27, 2017): 2946–2951, https://doi.org/10.1073/pnas.1617394114.

8. For an interesting discussion of the difference between experience and evidence, see Neil Burrows, "Conflicting Evidence: Prescribed Burning—

When 'Evidence' Is Not the Reality," keynote address, Australasian Fire Authorities Council Conference, Perth, Western Australia, September 5, 2018, available at www.researchgate.net/publication/327622300_Conflicting_evidence_prescribed_burning_-_when_%27evidence%27_is_not_reality.

9. See J. Russell-Smith, P. Whitehead, and P. Cooke, eds., "The West Arnhem Land Fire Abatement (WALFA) Project: the Institutional Environment and Its Implications," in *Culture, Ecology, and Economy of Fire Management in North Australian Savannas: Rekindling the Wurrk Tradition* (Tropical Savannas Cooperative Research Centre, 2009), 287–312.

10. For additional reading on each of these places, see the bibliographies in my essays in Stephen J. Pyne, *To the Last Smoke* (Tucson: University of Arizona Press, 2016): "The Mogollons: After the West Was Won," in *The Southwest: A Fire Survey*, vol. 5 of *To the Last Smoke*, 22–33; "Vignettes of Primitive America," in *California: A Fire Survey*, vol. 2 of *To the Last Smoke*, 167–176; and "Fire's Call of the Wild," in *The Northwest: A Fire Survey*, vol. 3 of *To the Last Smoke*, 33–43. The larger context for policy reform and its translation into practice is available in Stephen J. Pyne, *Between Two Fires: A Fire History of Contemporary America* (Tucson: University of Arizona Press, 2015).

11. The Yellowstone fires generated an expansive response; for a brief introduction to policy consequences, see Ron Wakimoto, "National Fire Management Policy," *Journal of Forestry* (October 15, 1990). There are many accounts of Cerro Grande, but the most neutral is Barry T. Hill, *Fire Management: Lessons Learned from the Cerro Grande (Los Alamos) Fire and Actions Needed to Reduce Risk*, GAO/T-RCED-00-273 (Washington, DC: US Government Accounting Office, 2000). On the Lower North Fork, see William Bass et al., "Lower North Fork Prescribed Fire: Prescribed Fire Review," report to the Colorado Department of Natural Resources (April 13, 2012), www.colorado.gov/pacific/sites/default/files/12Wildfire%20FireReview.pdf.

12. On field science challenging simulation science, see Burrows, "Conflicting Evidence."

Epilogue

1. This paragraph and the two that follow it, with minor edits, quote from the essay "Old Fire, New Fire," in Stephen J. Pyne, *Smokechasing* (Tucson: University of Arizona Press, 2003), 46–47. The most original source for the

ceremony is Fray Bernardino de Sahagun, "Florentine Codex: General History of the Things of New Spain," in Arthur J. O. Anderson and Charles E. Dibble, trans. and eds., *Book 7: The Sun, Moon and Stars, and the Binding of the Years*, Monographs of the School of American Research, No. 14, Part 8 (Santa Fe, New Mexico, 1953).

Bibliographic Essay

Since this book is an interpretive essay, or argued analogy, not a monograph, I have trimmed citations and source readings to a minimum. For further documentation, I refer readers to my *Fire: A Brief History,* 2nd ed. (University of Washington Press, 2019) and the chapters I contribute to Andrew Scott et al., *Fire on Earth: An Introduction* (Wiley-Blackwell, 2013). Behind these distillations lie major fire histories that I have written for Australia (*Burning Bush,* University of Washington Press, 1998), Canada (*Awful Splendour,* UBC Press, 2007), Europe including Russia (*Vestal Fire,* University of Washington Press, 2000), the United States (*Fire in America,* University of Washington Press, 1997; *Between Two Fires,* University of Arizona Press, 2015; *To the Last Smoke,* University of Arizona Press, 2020), and a score of short surveys on other countries.

I come from a tradition that deals preferentially with books, while science comes from a tradition of articles. Books can offer greater narratives, broader syntheses, and more examples; review articles are useful for surveying the status of a field. In what follows, I emphasize mostly books, supplemented by some recent scientific summaries, not so much for the ever-elusive general reader as for the generalist interested in how various disciplines understand the topic.

On the Pleistocene, I found particularly useful R.C.L. Wilson, S.A. Drury, and J.L. Chapman, *The Great Ice Age: Climate Change and Life* (Routledge, 2000); Clifford Embleton and Cuchlaine A.M. King, *Glacial Geomorphology,* 2nd ed., and *Periglacial Geomorphology,* 2nd ed. (John Wiley, 1975); and E.C. Pielou, *After the Ice Age: The Return of Life to Glaciated North America* (University of Chicago Press, 1991). A model of popular explanation, John

Imbrie and Katherine Palmer Imbrie, *Ice Ages: Solving the Mystery* (Harvard University Press, 1978), helped unravel the various contributing causes. For understanding the climate dynamics and their segue into the Holocene, William F. Ruddiman, *Plows, Plagues and Petroleum: How Humans Took Control of Climate* (Princeton University Press, 2005), was both critical and accessible.

On the history of fossil fuels, I found Vaclav Smil, *Energy in World History* (Westview Press, 1994), wonderful for historical context and J. R. McNeill, *Something New Under the Sun: An Environmental History of the Twentieth-Century World* (Norton, 2000), insightful for tracing the endless tendrils of a fossil fuel civilization. For competing concepts of the Anthropocene, Erle C. Ellis, *Anthropocene: A Very Short Introduction* (Oxford University Press, 2018), was especially useful.

And fire, in all its complexities? A bibliography would be endless, and over the past 25 years the literature has grown exponentially. The nonprofit Fire Research Institute has the most comprehensive collection for landscape fire, with monthly updates issued as Current Titles in Wildland Fire. Its database (172,000 items as of August 2020) can be accessed online at www.fireresearchinstitute.org. For placing industrial combustion within the very long history of fire and the controversy over fossil charcoal, there is Andrew Scott's very accessible extended essay, *Burning Planet: The Story of Fire through Time* (Oxford University Press, 2018).

Fire behavior, fire ecology, fire weather, fire history—all have books and review articles, usually by discipline; each of the major fire biomes has books or conference proceedings that summarize the fast-changing states of knowledge. Among the recent best are Jon E. Keeley et al., *Fire in Mediterranean Ecosystems: Ecology, Evolution and Management* (Cambridge University Press, 2012); Richard Cowling, ed., *The Ecology of Fynbos: Nutrients, Fire and Diversity* (Oxford University Press, 1992); Ross A. Bradstock, A. Malcolm Gill, and Richard J. Williams, eds., *Flammable Australia: Fire Regimes, Biodiversity and Ecosystems in a Changing World* (CSIRO, 2012); Mark A. Cochrane, *Tropical Fire Ecology: Climate Change, Land Use, and Ecosystem Dynamics* (Springer, 2009); and Jan W. Van Wagtendonk et al., eds., *Fire in California's Ecosystems*, 2nd ed. (University of California Press, 2018). For Spanish speakers, see Dante Rodriguez Trejo, *Incendios de vegetacion: Su ecologia, manejo e historia*, 2 vols. (Colegio de Postgraduados, 2014, 2015). The six-volume series *Wildland Fire in Ecosys-*

tems, published by the US Forest Service, offers summaries for fire effects on soil, water, air, flora, fauna, invasive species, and archaeological sites.

From various disciplines now interested in fire, see the still valuable William J. Bond and Brian W. van Wilgen, *Fire and Plants* (Chapman and Hall, 1996). A general-audience introduction from the perspective of a chemist and urban fire specialist is John W. Lyons, *Fire* (Scientific American Books, 1985). Still unrivaled as a popular prelude is Mark J. Schroeder and Charles C. Buck, *Fire Weather . . . A Guide for the Application of Meteorological Information to Forest Fire Control Operations,* Agriculture Handbook 360 (US Forest Service, 1970). For humanity's relationship with fire, as viewed by a perceptive European sociologist, see Johan Goudsblom, *Fire and Civilization* (Penguin Press, 1992). On the growing concern over smoke and health, see Fay H. Johnston, Shannon Melody, and David M. J. S. Bowman, "The Pyrohealth Transition: How Combustion Emissions Have Shaped Health through Human History," *Philosophical Transactions of the Royal Society B* 371: 20150173, https://doi .org/10.1098/rstb.2015.0173. From ethnobotanists comes Cynthia T. Fowler and James R. Welch, eds., *Fire Otherwise: Ethnobiology of Burning for a Changing World* (University of Utah Press, 2018); and from paleoanthropologists, Richard Wrangham, *Catching Fire: How Cooking Made Us Human* (Basic Books, 2009). I would be remiss not to include the still-invaluable Harley H. Bartlett, *Fire in Relation to Primitive Agriculture and Grazing in the Tropics: Annotated Bibliography* (University of Michigan Botanical Gardens, June 1955). And I can't refuse two short publications by The Nature Conservancy that explain fire and humans with remarkable clarity: A. J. Shlisky et al., *Fire, Ecosystems, and People: Threats and Strategies for Global Biodiversity Conservation* (2007) and Ronald L. Myers, *Living with Fire: Sustaining Ecosystems and Livelihoods through Integrated Fire Management* (2006).

Some recent articles are helpful in placing fire within existing disciplinary understandings: David M. J. S. Bowman et al., "Vegetation Fires in the Anthropocene," *Nature Reviews Earth and Environment* (August 18, 2020); Jennifer R. Marlon, "What the Past Can Say about the Present and Future of Fire," *Quaternary Research,* https://doi.org/10.1017/qua.2020.48; N. Andela et al., "A Human-Driven Decline in Global Burned Area," *Science* 356 (2017): 1356–1362; International Savanna Fire Management Initiative, "The Global Potential of Indigenous Fire Management: Findings of the Regional Feasibility Assessments" (United

Nations University, 2015); Christopher I. Roos et al., "Living on a Flammable Planet: Interdisciplinary, Cross-Scalar and Varied Cultural Lessons, Prospects and Challenges," *Philosophical Transactions of the Royal Society B* 371: 20150469, http://dx.doi.org/10.1098/rstb.2015.0469; and Meg A. Krawchuk et al., "Global Pyrogeography: The Current and Future Distribution of Wildfire," *PLOS ONE* 4(4) (April 2009), e5102. For an example of big data applied to pyrogeography, see Nathan Mietkiewicz et al., "In the Line of Fire: Consequences of Human-Ignited Wildfires to Homes in the U.S. (1992–2015)," *Fire* 3 (2020): 50, https://doi.org/10.3390/fire3030050. And an attempt to survey the field for ecologists, Kendra K. McLauchlan et al., "Fire as a Fundamental Ecological Process: Research Advances and Frontiers," *Journal of Ecology* 108 (2020): 2047–2069.

Not least, for its scope and sustained commitment, there is the Global Fire Monitoring Center (https://gfmc.online), with its rich depositories of news, institutional arrangements, country reports, and general compendium of landscape fire management at a planetary scale.

The above roster is a sampler. The list of potential candidates is bottomless, and for every one listed there are a dozen members of the fire community annoyed that their special contributions have been ignored. The good news is that, as with fire, so with fire research: many points of entry exist. Whatever one's particular passion, it is likely that its specialty discipline has been touched by fire.

Index

model, 134; as management strategy, 132–36; in western US, 134–35. *See also* aboriginal fire; agricultural fire

prescribed natural fire, 133–34

Promethean narrative, 149–50

Prometheus, 84

pyric transition, 84–89, 106–7; concept of, 89–92

Pyrocene: as analogy, 5–6; concept of, 3–6; as narrative, 146–47, 149–50; paradoxes of, 3, 147; various meanings of, 146

pyrocumulus cloud, 107–8

pyrotechnics, 74–77; biological, 76–77; biological versus physical, 76–77; physical, 74–75

Quaternary, 36, 46–49. *See also* Holocene; Pleistocene

sapiens, 27, 44–45, 49–52, 57, 64, 77, 119, 124–25

second-fire, 28, 55, 66, 80, 89, 91–92, 96, 106, 112, 126, 128–29; characteristics of, 56–58; concept of, 4–5, 56–58; contrast with third-fire, 5, 89, 92, 96, 106, 126, 128, 129, 132, *plates 2, 6, 9*; influence on climate, 77–80; interaction with first-fire, 65–66, 89; as management strategy, 132–36; as missing, 93, 116–17

Shelley, Mary, 84

Shelley, Percy, 84

Show, S. B., 104

slow combustion, 11–13, 15, 27, 71

smoke: ecology of, 17, 19–20; health hazard of, 32, 95, 121; in third-fire, 95–97, 129; in 2019–20 fire season, 1–3, 116; uses of, 59, 64, 75

St. Francis of Assisi, 149

substitute tame fire for wild fire, as management strategy, 132–36

swidden, 69–71, 130; Finnish terms for, 70; invention of term, 70. *See also* agricultural fire; second-fire

Tallgrass Prairie Preserve (Oklahoma), *plates 2, 3*

Tall Timbers Research Station (Florida), 31, 133

third-fire, 81, 89, 91, 93, 106, 121, 126, 128–29, 132; concept of, 4–6, 88; contrast to second-fire, 85–89; effect on fallow, 98; global consequences, 91–92, 106; impact on built landscapes, 94–95; impact on colonization, 100; impact on rural landscapes, 96–98; impact on wildlands, 98–105; influence on climate, 80–81, 105–7; interactions with first- and second-fire, 4, 55, 66, 89, 112, 128, *plates 1, 3, 4, 6, 9*; relationship to nature preserves, 99, 104–5. *See also* industrial combustion; lithic landscapes; pyric transition

three fires: concept of, 4–6; interactions among, 4, 55, 66, 89, *103fig.*, 105–8, 109–16, 127–29, *plates 1, 3, 4, 6, 9. See also* first-fire; second-fire; third-fire

traditional ecological knowledge, 142

2019–20 fire season, 1–2, 112–16

Founded in 1893,
UNIVERSITY OF CALIFORNIA PRESS
publishes bold, progressive books and journals
on topics in the arts, humanities, social sciences,
and natural sciences—with a focus on social
justice issues—that inspire thought and action
among readers worldwide.

The UC PRESS FOUNDATION
raises funds to uphold the press's vital role
as an independent, nonprofit publisher, and
receives philanthropic support from a wide
range of individuals and institutions—and from
committed readers like you. To learn more, visit
ucpress.edu/supportus.